W9-AFS-640

Introduction to the Study of Animal Populations

Introduction to the Study of Animal Populations

H. G. ANDREWARTHA

Department of Zoology,
The University of Adelaide,
Adelaide, South Australia

SECOND EDITION

THE UNIVERSITY OF CHICAGO PRESS

The University of Chicago Press, Chicago 60637
*Methuen & Co. Ltd, London E.C.*4

Published 1961. *Second Edition* 1971
Printed in Great Britain

International Standard Book Number: 0-226-02029-0
Library of Congress Catalog Card Number: 73-135741

Contents

7. *Components of environment; malentities*

8. *Components of environment; more about the ecological web*

9. *Theory; the numbers of animals in natural populations*

Preface to the First Edition

My interest in animal ecology was first aroused by reading Elton's *Animal Ecology*. His definition of the scope of ecology which I quote in section 1.0 is still the best that I have met.

Notwithstanding the breadth of Elton's original definition the tendency of modern theoretical ecology has been to stress the interrelationships between animals at the expense of other components of environment which may also influence the 'distribution and numbers of animals in nature'. On the other hand the published works of practising ecologists, in such fields as economic entomology or the conservation of wildlife, have provided a wealth of information about how the distribution and numbers of animals in nature are in fact determined. Andrewartha and Birch attempted to systematize this information in *The Distribution and Abundance of Animals*.

The course in animal ecology that I teach in Adelaide makes use of Andrewartha and Birch's theory of environment. But I have felt the need for a compact text that the students may use. Also when the theory of animal ecology is taught this way it seems natural to join it to a practical course that is quantitative and largely experimental. In writing this book I have had these two objectives in mind. Part I gives a compact exposition of Andrewartha and Birch's theory of environment; Part II is a manual of practical exercises that arise from and are related to this theory.

Section 9.3 is addressed primarily to teachers and advanced students. If ecology is to be taught narrowly with the emphasis on the relationships between the animals in communities the issue of 'density-dependent factors' may not have to be faced seriously. But if ecology is to be taught broadly giving proper emphasis to all aspects of environment then the teacher must decide for himself what part, if any, the concept of density-dependent factors may play in a general theory of environment. I hope that section 9.3 may help him to reach a critical decision. The undergraduate should read section 9.33 as an introduction to Part II even if he does not read the rest of section 9.3 as an extension of Part I.

The experiments which are described in Part II may be divided into those dealing with methods for measuring dispersal, distribution and density and those dealing with physiological and behavioural responses of animals to particular components of environment such as temperature, food, other animals, etc. The theoretical background for the experiments in Part II is discussed in Part I, in relation to its proper place in the general theory. This arrangement has allowed me to set out the experiments in Part II with economy, as in a laboratory notebook; with respect to each group of experiments I have indicated the sections in Part I that are relevant; and then each experiment is set out under stereotyped headings viz. – (*a*) Purpose, (*b*) Material, (*c*) Apparatus, (*d*) Method, (*e*) Analysis of results. This arrangement also has the advantage of weaving theory and practice closely together.

Ecology, by its very nature, deals with populations of living animals and it becomes a dull subject unless it can be related to natural populations. Ideally the methods of estimating density, dispersal and distribution should be taught with populations that are living naturally in the field. I do this so far as it is practicable. But, in some instances the methods can also be taught by means of 'laboratory models', as, for example, in experiment 10.25 where we estimate the density of a population of *Tribolium* by the method of capture, marking, release and recapture. The laboratory model has the advantage that it can be done indoors during bad weather or during winter. Also it can usually be relied upon to give results that can be analysed.

Most of the animals that are available for study and the sorts of places where they live are characteristic of particular climatic or zoogeographic regions. This poses the problem of how to write a manual of practical animal ecology that might be useful outside the region where it was compiled. But ecological principles and techniques can be demonstrated with a wide variety of experimental animals, and I have chosen species that are cosmopolitan, or at least, are similar to those that are distributed widely in other continents, e.g. grain beetles, white cabbage butterfly the garden snail and the house mouse. But, apart from these, there must be, in any region, many local species that are suitable for quantitative ecological experiments.

I have been helped not only to think but also to write more clearly by many long discussions with Dr T. O. Browning. My wife has read and criticized the text, and helped with the preparation of indexes. Many of the illustrations have been taken from *The Distribution and Abundance of Animals*, but a number are original. They were drawn by Mrs P. E. Madge. I am grateful to Mr G. Wilkinson and Dr A. James for help with statistics, especially with the method for estimating dispersal.

H. G. A.

Adelaide, 1959

Preface to the Second Edition

In revising this book I have tried to bring the theory of environment up to date in the light of certain important criticisms that have appeared since 1961, especially in papers by T. O. Browning and D. A. Maelzer, and in the light of experience gained while using the book as a text for an undergraduate course in population ecology in the University of Adelaide. As a consequence the order in which the argument is presented has been altered. Some new material has been introduced to expand the discussion of certain topics, especially resources, pathogens, aggressors and territorial behaviour. But the general approach to the subject and the general theory remains very much the same as in the first edition.

I am grateful to Professor F. Fenner and Dr F. N. Ratcliffe and to Cambridge University Press for permission to reproduce Fig. 5.04; to Professor D. O. Chitty and the Ecological Society of Australia for permission to reproduce Fig. 5.05 (with minor modifications); Fig. 3.03 has been modified from a figure in a paper by H. G. Andrewartha and T. O. Browning first published in the *Journal of Experimental Biology*.

Adelaide, 1970 H. G. A.

PART I

Theory

I The history and scope of ecology

I.O INTRODUCTION

In their *Principles of Animal Ecology* Allee *et al.* (1949) wrote a long introductory chapter on the history of the subject. They went back to the writings of Theophrastus, who was a friend of Aristotle, and who was interested in the sorts of animals and plants that could usually be found living together in 'communities'. Allee *et al.* also found much that was relevant to ecology in the writings of Pliny and other Romans, and, of course, in the writings of scholars who lived in various parts of Europe after the Renaissance.

In more recent times the great French naturalists Réaumur (1663–1757) and Buffon (1707–88), and the German, Ernst Haeckel, who coined the word 'Oekologie' in 1869, all helped to build the foundations of the modern science.

The publication in 1927 of Elton's book *Animal Ecology* greatly influenced the development of modern ecology. The fashion in biology had previously been set by the publication in 1859 of *The Origin of Species* and the great controversies which followed. Taxonomy, comparative anatomy, evolution and phylogeny had been the chief interest of biologists for more than half a century. Elton, protesting against this perspective, wrote:

> In solving ecological problems we are concerned with *what animals do* in their capacity as whole, living animals, not as dead animals or as a series of parts of animals. We have next to study the circumstances under which they do these things, and, most important of all, the limiting factors which prevent them from doing certain other things. By solving these questions it is possible to discover the reasons for the *distribution and numbers of animals in nature.*

Elton's great contribution was to point out the importance of studying the distribution and abundance of animals in nature and to demarcate this as a proper field for ecology. He went on to say that a good way to go about solving these problems was to study the relationships between animals living together in 'communities'. He thought of a community as the complex of animals that are usually found living together in a 'habitat'. The meaning of 'habitat' was made clear in Elton's (1949) essay on 'population interspersion'. A habitat is defined as an area that seems to possess a certain uniformity with respect to physiography, vegetation or some other quality that the ecologist decides is important (or easily recognized). He decides on the limits of the habitat arbitrarily, and in advance, as a first step towards his study of the community.

The sort of relationships which Elton emphasized may be defined by explaining the concept of 'niche'. Elton (1927) said that all communities of plants and animals have a similar ground-plan; they all have their herbivores, carnivores and scavengers. In a wood there may be certain caterpillars which eat the leaves of trees, foxes which hunt rabbits and mice, beetles which eat springtails and so on. 'It is convenient to have some term to describe the status of an animal in its community, to indicate what it is *doing* and not merely what it looks like, and the term used is "niche". . . . The "niche" of an animal means its place in the biotic environment, *its relations to food and enemies*.' Thus the caterpillar and the mouse broadly occupy similar niches because they eat plants; the fox and the ladybird also fit broadly into similar niches because they both eat other animals in the community. But we need to have more information than this about 'niches'.

For example, with respect to the 'niche' of the fox, it is not sufficient merely to know that 'foxes eat rabbits' and 'foxes can flourish on a diet of rabbit'. In addition we need to inquire 'how many foxes are there?' and 'what are their chances of getting enough food?'

So we may conveniently recognize three levels of complexity in the laws of ecology. Firstly, there are the laws governing the physiology and behaviour of individuals in relation to their environments; secondly, there are the laws governing the numbers of animals in relation to the areas that they inhabit; and thirdly, there are the laws governing *communities*. A community may be thought of as an interacting population of plants and animals. The laws governing communities explain the interlocking network of niches which tie the different species together as a community. One such network links the organisms in a hierarchy according to how they get their food. The green plants, because they contain chlorophyll, can synthesize organic matter from carbon dioxide and water, transforming the energy of sunlight to the energy of chemical bonds. Some of this chemical energy is used by the plant as it grows; some is stored in its tissues which are subsequently eaten by herbivores or decomposed by saprophytes. None of the herbivores, carnivores or saprophytes in the community can make its own food in this way. So the plants are called *autotrophs* and all the rest *heterotrophs*. The existence of virtually the whole community depends on the passage of the primary stock of energy, accumulated by the autotrophs, through the successive trophic levels of the community, as herbivore eats plant, carnivore eats herbivore or carnivore, and saprophyte consumes the dead remains of plant, herbivore, carnivore or saprophyte. Eventually, perhaps after many passages, the energy is dissipated as heat, but meantime is being continuously replenished from sunlight. The branch of ecology that studies the flow of energy through a community has been given the special name *community energetics*, and the whole system including the inflow and outflow of energy and the circulation of matter, oxygen, nitrogen, phosphorus, calcium, etc. has been called an *ecosystem*.

Because most books about ecology pay a lot of attention to communities and ecosystems the laws governing them are well developed and generally agreed upon. In addition to the simple idea of energy flowing through the successive trophic levels the study of ecosystems has given rise to a number of interesting ideas about the ontogeny of communities and especially about the limitations imposed on the development of communities. But the laws governing ecosystems have not led, nor might they be expected to lead, directly to an explanation of the distribution and abundance of animals in nature which Elton saw as the essence of animal ecology. When a naturalist contemplates organic nature inevitably he sees a community or part of one; if at first he sees only a small segment he has only to enlarge the area of his observation to encompass a functional community because that is the way that organisms are organized in nature. But to recognize a population a naturalist must deliberately concentrate his thoughts on the individuals of one species; moreover he must usually make the further decision to consider only those members of the species that live within an arbitrarily restricted area because it is not often practicable to contemplate all the members of a species over its whole distribution. So the idea of 'population' is more abstract than the idea of 'community' or of 'ecosystem'. Nevertheless there is one important context in which the idea of population seems more real than the idea of ecosystem. For example, the farmer who wishes to get the maximum harvest from his crop of wheat or from the flock of sheep that graze his pastures may wish that there were fewer thistles among his wheat or fewer caterpillars eating his pasture. But he would merely confound the issue, nor would he advance one whit towards the solution of his problem, if he were to measure the wheat and the thistles together as the first trophic level of an ecosystem or to lump the sheep and the caterpillars together as the second trophic level in another ecosystem. So also with most of the problems that characteristically confront farmers, graziers, fishermen, foresters and even sociologists and politicians who contemplate the population explosion that threatens disaster for our own species. These are all problems in population ecology which needs a theory of its own.

1.1 POPULATION ECOLOGY: THE STUDY OF THE DISTRIBUTION AND ABUNDANCE OF ANIMALS

Some time ago I spent a number of years studying the ecology of *Thrips imaginis* which is a small winged insect that lives and breeds in the flowers of a wide variety of garden plants, weeds, fruit trees and native shrubs around Adelaide, South Australia. One reason for choosing to work on this particular species was that it had recently (during the spring of 1931) multiplied to such enormous numbers that it had caused what was virtually a complete failure of the apple crops of South Australia, Victoria and New South Wales. The apple-growing districts in South Australia are separated from those in Victoria by about 500 miles, and those in New South Wales are still farther away. No one counted the numbers of *Thrips imaginis* in apple orchards very thoroughly that year. But we know now that it takes about 40 thrips to destroy one blossom.

Starting with this figure, and taking into account the number of flowers that an apple tree might bear, I have guessed that there might have been about 20 million thrips per acre in most of the apple orchards in south-eastern Australia in 1931; they were also present in comparable numbers outside the orchards in gardens, fields and elsewhere.

We set out to explain the unusual abundance of *Thrips imaginis* during 1931; of course this meant that we had to try to explain the abundance of *Thrips imaginis* at all times – not merely during such extraordinary plagues. It seemed reasonable to start with the hypothesis that the plague had been caused primarily by weather, because we could think of no other component of environment that would operate over such a vast area. So we set out to study all the ways that weather might influence the fecundity, speed of development and length of life – and hence the abundance – of *Thrips imaginis*.

From the beginning it was clear that we would have nothing to explain unless we knew how many *Thrips imaginis* were present in our study-area. So we began counting the thrips in the large garden where we were working, and we continued to count them every day for 14 years. It was these counts that formed the fundamental basis around which we planned all our concomitant studies of behaviour, physiology,

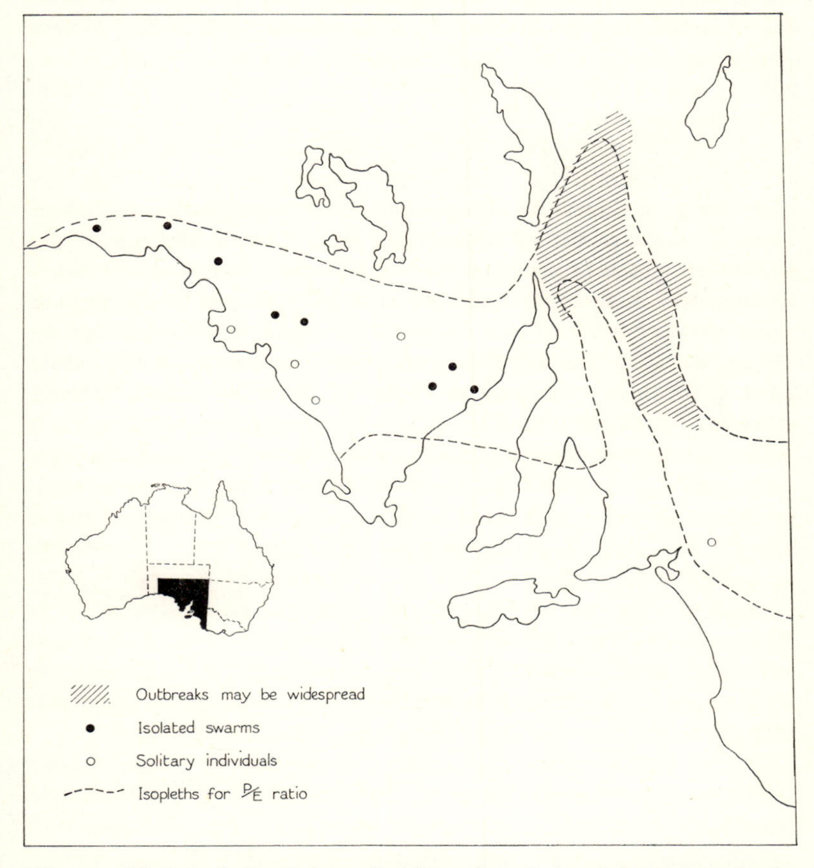

Fig. 1.01. The area in South Australia where *Austroicetes cruciata* may maintain a dense population when the weather is favourable. The isopleths for P/E ratio define the boundaries of a climatic zone to which the 'grasshopper belt' is largely confined. (After Andrewartha, 1944.)

meteorology, etc. – in a word the ecology of *Thrips imaginis*. A few years later I began to study the ecology of a grasshopper, *Austroicetes cruciata*, which, at that time, was causing a lot of damage to pastures and crops of wheat in one part of South Australia. During the next four or five years I found that swarms of this grasshopper occurred each year in a large but nevertheless well-defined zone (Fig. 1.01). I studied the old records and found that during the preceding 50 years there had been about seven outbreaks, and on every occasion the grasshoppers seemed to have been restricted to the same area that I had found them in during 1935–40. In this instance the most interesting scientific problem seemed to be to find an explanation for the *distribution* of this grasshopper, and to discover the reasons for the sharply defined boundaries to the area in which it was able to multiply to plague numbers. We discovered that the boundaries of the distribution were closely related to climate (Andrewartha, 1944), and it soon became clear that the components of climate which determined the limits of the distribution were the same as the components of weather that determined the abundance of the grasshoppers inside the area of their distribution. In other words, there were not two separate problems of distribution and abundance to be explained, but only two aspects of the same problem. As our studies of *Thrips imaginis* progressed we saw that the same was true for this species too – as I have explained elsewhere (Andrewartha and Birch, 1954, p. 582).

In order to generalize this point I have drawn Figs. 1.02 and 1.03.

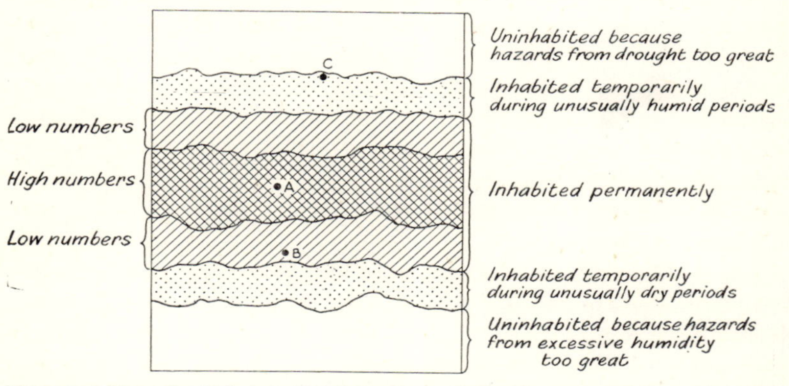

Fig. 1.02. Hypothetical map of the distribution of any animal based on the real distribution of the grasshopper *Austroicetes cruciata* (Fig. 1.01). For further explanation see text. The localities A, B, C also feature in Fig. 1.03. (After Andrewartha and Birch, 1954.)

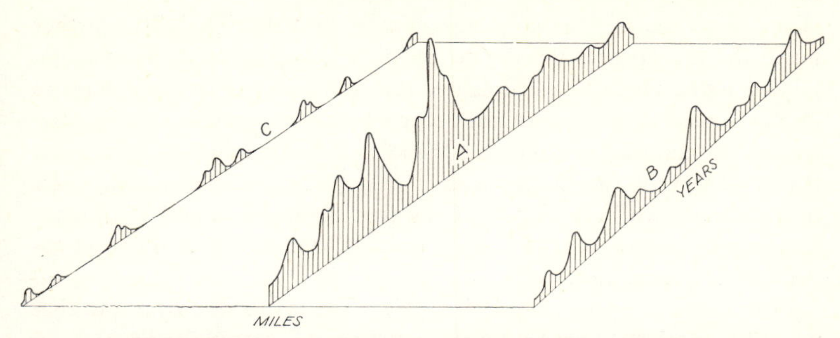

Fig. 1.03. Fluctuations in the abundance of the hypothetical animals that live at localities A, B, C in Fig. 1.02. For further explanation see text. (After Andrewartha and Birch, 1954.)

They are based on the observed distribution of *A. cruciata*, but they also serve to illustrate the relationship between distribution and abundance that holds generally for all animals. For *A. cruciata* we were able to map a zone inside which we could reasonably expect to find the grasshoppers whenever we looked for them. Although we could draw the boundaries of this zone fairly precisely for any one year, the boundaries were not found in precisely the same place from year to year. They fluctuated a little, chiefly in response to weather. During a run of wet years the southern extremes of the area shown in Fig. 1.01 tended to become too wet for the grasshoppers (they were likely to be killed by outbreaks of fungal disease) but were able to spread a little farther north towards the desert. Similarly, during a run of dry years the northern extremes were likely to be too dry (the grasshoppers were likely to die from lack of food), but they were able to spread a little farther south. This is the sequence of events that I have tried to summarize in Figs. 1.02 and 1.03. In Fig. 1.02 the emphasis is on the fluctuating shape and size of the distribution; in Fig. 1.03 I have emphasized fluctuations in abundance, not only from year to year but also from place to place; the two diagrams are linked by the localities A, B and C which are shown in both.

The grasshoppers were usually more abundant in the central, more favourable, parts of the distribution (locality A), where the weather was usually neither too wet nor too dry. They were less abundant towards the margins of the distribution where the weather was more often too wet or too dry. The weather was never quite wet enough for long enough at B to reduce the population to zero; at C the weather was, from time to time, dry enough to wipe out the entire population; but during runs of wet years the area might be re-colonized temporarily. A little farther north the climate was too dry to permit even temporary colonization. Thus I found when I had explained the abundance of the grasshoppers in all the places that they might occur I had also explained their distribution. It seems logical to define the idea of distribution by saying that the boundary of the distribution is that place where the abundance decreases to zero. This is why I say that distribution and abundance should not be treated as two separate ideas, but merely as different aspects of the same idea.

With both *T. imaginis* and *A. cruciata* the systematic counting that we did formed the essential basis for the whole study. The problem of estimating the numbers in a natural population is basic for most field work in ecology. The emphasis may shift to different aspects: with *T. imaginis* the emphasis was chiefly on fluctuations in abundance from year to year; with *A. cruciata* this was important also, but the emphasis was rather more on their distribution; on another occasion one might wish to compare the relative abundance of the animals from place to place in some small part of their distribution. Whatever the aspect of ecology that is being emphasized, the first essential for good work is to know how many animals there are in the population that you are studying. The population is an abstraction because it has been arbitrarily defined as belonging to a particular area which is smaller (usually very much smaller) than the entire distribution of the species. The area is usually chosen for its convenience as a 'study-area'; usually it is chosen large enough to ensure that its population does not become extinct during the course of the study; but sometimes small 'local areas' are chosen deliberately, with the expectation that the 'local population' will become extinct during the study (chapter 9); rarely is the study-area surrounded by ecological barriers that might prevent migration into or out of it. It is well to remember these limitations to the definitions of 'population' in order to guard against the temptation to attribute more reality to the idea than it deserves. The meaning to be attributed to 'density,' and some techniques for measuring the density of natural populations are discussed in chapter 10.

1.2 THE BROAD BASES FOR POPULATION ECOLOGY

Population ecology deals with the numbers of animals that can be counted or estimated in natural populations; and the essence of the science is to explain these numbers. That is why ecologists need to learn at least the fundamentals of statistics, not only so that they can use the routine statistical methods to analyse the results of their experiments, but also in order that they may learn to think about their work in terms of probabilities.

A knowledge of the behaviour of animals and their physiology is also a necessary part of the background for studying ecology.

I hope that the ways in which these skills are used in ecology will become clear in the course of this book. It is not so easy to write about the need for the arts of the naturalist – although this is perhaps the most important quality of all. It is best conveyed by the teacher, verbally and by precept. There is a passage from Ford (1945) which makes the point nicely.

We have three butterflies which are limited by geological considerations, being inhabitants only of chalk downs or limestone hills in south and central England, and they may reach the shore where such formations break in cliffs to the sea. These are the silver-spotted skipper, *Hesperia comma*, the chalk-hill Blue, *Lysandra coridon* and the Adonis Blue, *L. bellargus*. . . . The two latter insects are further restricted by the distribution of their food plant, the Horseshoe Vetch, *Hippocrepis comosa*, and possibly by a sufficiency of ants to guard them. Yet any of the three may be absent from a hillside which seems to possess all the qualifications which they need, even though they may occur elsewhere in the immediate neighbourhood. This more subtle type of preference is one which entomologists constantly encounter, and a detailed analysis of it is much needed. A collector who is a careful observer is often able to examine a terrain and decide, intuitively as it were, whether a given butterfly will be found there, and that rare being, the really accomplished naturalist, will nearly always be right. Of course he reaches his conclusions by a synthesis, subconscious as well as conscious, of the varied characteristics of the spot weighed up with great experience; but this is a work of art rather than of science, and we would gladly know the components which make such predictions possible.

Glover *et al.* (1955) described how this 'naturalist's art' was used in a campaign to exterminate the tsetse fly *Glossina morsitans* from 280 square miles of bush in northern Rhodesia. It was known that *G. morsitans* rarely hunts for food in dense scrub but only in places where the vegetation is open. So Glover and his colleagues set out to prevent bushfires and do everything else that they could to make the bush grow more quickly. In this they were largely successful. At the end of 10 years the undergrowth was considerably thicker, but there was not much difference in the density of the taller vegetation. By the end of five years the flies (taken in a routine sample) had decreased from 5,915 to 4,739 but there was no indication that they were likely to be exterminated from the area by these methods.

About this time it was noticed that *Glossina* tended to be slightly more numerous in certain small valleys which were known locally as 'kasengas'. Glover *et al.* described one characteristic kasenga as follows.

'It starts between high hills as a narrow but flat-bottomed cleft a little under two miles long: after two miles it acquires a definite water-course and becomes a rocky ravine. The upper flat-bottomed section supports a fair growth of trees, and the adjacent hillsides are also well wooded with *Brachystegia:* the trees in the valley are mainly *Terminalia* and *Combretum.*' On the assumption that the kasengas were important for the tsetse flies, Glover *et al.* decided to clear them. The trees along the small valleys and along the edges of the surrounding woodland were felled. The clearing was started in 1941 and finished in 1944, altogether 2·2% of the total area was cleared.

The results were immediate and spectacular. The number of flies caught in a routine sample which had dropped by about 35% during the five years of 'fire exclusion', dropped, during the five years after clearing of kasengas began, from 2,750 to 40; by 1950 Glover *et al.* were satisfied that *G. morsitans* had been extermined from the whole area of 280 square miles.

The final extinction of the population was probably hastened by failure of the sexes to meet, and there was some suggestion that right from the beginning the chief outcome of clearing the kasengas was to reduce the likelihood that the flies would mate. Glover *et al.* commented on this as follows.

In 1942, when fly numbers had already been greatly reduced, an uninseminated though non-teneral female fly was captured in the Isoka area; such an observation had never before been made by the Tanganyika Department of Tsetse Research, who had invariably found that all wild females which had taken their first meal proved on dissection to be inseminated. In the following year, 12 non-teneral females out of 136 (9% were uninseminated; in 1944 the figure was 20% of 152 and in 1945 it was 29% of 42 dissected. These findings strengthened our belief that there was later no breeding population of flies within the area, since as early as 1942, when numbers were relatively high, the sexes were beginning to fail to encounter each other, and this effect increased yearly to 1945, after which numbers became extremely low and it seems certain that the great majority of females must then have remained virgin (see also chapter IV).

It was suggested as a result of these observations that the kasengas might have been important because the flies sought them out and stayed in them until they had mated; but this suggestion was not verified. The authors conclude their paper with the comment: 'It must be confessed that we do not know what vital role these valleys played in the ecology of the flies; we only know that cutting down the trees in and immediately around them put them out of action.'

The ecology of three species of deermice, *Peromyscus maniculatus, P. californicus* and *P. truei,* was studied in an area of 28 square miles near Berkeley, California, by McCabe and Blanchard (1950). The area comprised a ridge of hills dissected by many valleys; there was a diversity of aspects and soils supporting woodland, grassland, a dense scrub called 'chaparral' and several other types of vegetation. This study was a success because McCabe and Blanchard, exercising the 'naturalist's art' mentioned by Ford, were able to recognize the sorts of places in which the deermice would live.

The deermice were found only along the margins of the chaparral. None was trapped right inside the chaparral or inside any other type of vegetation. But wherever the chaparral met woodland or grassland, or along the edges of artificial clearings or roads through the chaparral, the mice were found.

The margins differed in their general *facies* and also in their suitability for the different species. The places most favoured by *P. maniculatus* were the moister slopes where a dense stand of chaparral gave way to a narrow band of giant herbs, mostly Umbelliferae with grassland behind them. The ground under the chaparral was bare except for a thin covering of dead leaves.

The other two species were most likely to be found, and their populations were most likely to be dense, along edges of the chaparral where it met woodland of *Pinus, Eucalyptus* or *Quercus*. It was characteristic of the places favoured by *P. truei* and *P. californicus* that the chaparral would be composed of a greater variety of species and its margins would be more sinuous than those that were favoured by *P. maniculatus*. Their requirements were more stringent than those of *P. maniculatus* because they were always very few except in places that suited them precisely, But *P. maniculatus* was almost ubiquitous, being present, if not abundant, wherever the chaparral manifested a well-defined edge.

Along some edges that suited *P. maniculatus* well this species made up 95% of the catch. In places that were more suitable for *P. truei* and *P. californicus* these species might make up 75% of the catch. No place was found where the population consisted exclusively of one species. So their requirements overlapped to some extent although each species had its own characteristic requirements of the places where it would live.

The ecology of the rabbit, *Oryctolagus cuniculus*, is probably better known than that of any other mammal, chiefly as a result of recent work in Australia by a team of ecologists in the C.S.I.R.O. Division of Wildlife Research. Rabbits live gregariously in clusters of underground burrows known as *warrens*. Soil and topography in relation to weather, vegetation and predators influence the distribution of warrens. Once a rabbit has lived in a warren, either as an adult that has been accepted into one of the resident social groups or as a juvenile that was born and reared there, it develops a strong attachment to that particular warren. Once the behaviour of the rabbit in relation to the place where it lives was appreciated, good progress was made with the study of its ecology.

Indeed it is generally true that each species of animal has particular, often narrow, requirements of the place where it will live. Ecologists need to learn the 'naturalist's art' of recognizing these requirements because this is the first and an essential step in studying its ecology. It is necessary to gain some intimacy with the animal's way of life before it is practicable to formulate hypotheses to explain its distribution and abundance.

1.3 HOW TO WRITE ABOUT ECOLOGY

It is an essential part of science that the scientist who has made a discovery should be able to tell his colleagues exactly what he did, how he did it and what are the conclusions to be drawn from his work. The clear, concise writing that is demanded of a scientist does not come naturally, but only after many trials and much practice. The young scientist should avoid the mistake of copying the jargon that characterizes far too much of the 'scientific literature'. He would be wiser to study the masters of non-scientific literature because much the same rules govern the writing of simple, lucid English whether you are describing the animals in a wood or the actors in a play. Even the choice of words need not be as different as some scientists would make it. I have gathered together, in the next few pages, several pieces of sound advice from literary men as well as from scientists on this important subject. Somerset Maugham (1948, p. 40), discussing the way that he set about attaining a style which would satisfy him, wrote:

On taking thought it seemed to me that I must aim at lucidity, simplicity and euphony. I have put these three qualities in the order of importance I assigned to them . . . You have only to go to the great philosophers to see that it is possible to express with lucidity the most subtle reflections. . . . There are two sorts of obscurity that you find in writers. One is due to negligence and the other to wilfulness. People often write obscurely because they have never taken the trouble to learn to write clearly. This sort of obscurity you find too often in modern philosophers, in men of science and even in literary critics. . . . Yet you will find in their works sentence after sentence that you must read twice in order to discover the sense. Often you can only guess at it, for the writers have evidently not said what they intended. . . . Some writers who do not think clearly are inclined to suppose that their thoughts have a significance greater than at first sight appears. It is flattering to believe that they are too profound to be expressed so clearly that all who run may read, and very naturally it does not occur to such writers that the fault is with their own minds, which have not the faculty of precise reflection. Here again the magic of the written word obtains. It is very easy to persuade oneself that a phrase that one does not quite understand may mean a great deal more than one realizes. From this there is only a little way to go to fall into the habit of setting down one's impressions in all their original vagueness.

Leeper (1941, 1952) commented on two faults which, he said, often make the writings of scientists difficult to understand.

'. . . I have become steadily more convinced that the abstract noun is most to blame for the common heaviness. The first step on the downward path is to write "solution takes place" instead of "the substance dissolves". The most important place is given to a noun instead of a verb, while the verb collapses to the emptiness of "take place" or "occur".'

Once the first downward step is taken and the abstract noun is accepted as the pivot-word of the sentence, the second step is easy. This is the use of sets of two or more nouns, only the last of which has the full status as a noun, while the preceding nouns qualify it. The double noun, of course, is part of our language. Some double nouns, like *boatrace* and *newspaper*, have even been accepted as single words, while many others are commonly used without offence, like *vapour pressure* and *spot test*. But a writer of any sensibility will use them sparingly, and will avoid double nouns which are not in everyday speech, such as the ugly *fertility decline* or *phosphate source*. Gowers (1948) has truly written 'This unpleasant practice is spreading fast and is corrupting the English language'. He quotes Dunsany's satiric remark that we may use the frequency of multiple nouns in a given passage for dating our *language decay progress*. A treble noun like this is always inexcusable, yet treble or even quadruple nouns are common in our journals.

'The vice of this habit is not that experts think it is ugly, but simply that it makes reading difficult. We do not know at first whether we are reading about *language*, *decay* or *progress*, and we must waste some seconds in reading the sentence over in order to solve the puzzle.'

Weiss (1950) mentioned a fault that may make the writings of a scientist not only difficult to comprehend but also misleading. There is a tendency, he said, to divest a phenomenon of its natural complexity by giving it a high-sounding name. Then the next step, which comes all too easily, is to slip into the habit of thinking that the high-sounding name describes or explains the nature of the phenomenon (section 6.111).

Orwell's (1946) five rules are a useful guide for scientists. I repeat them below, not quite literally.

(1) Never use a long word where a short one will do.
(2) If it is possible to cut out a word always cut it out.
(3) Never use the passive where you can use the active.
(4) Use technical words sparingly and correctly; never use a word that cannot be understood by a scientist working in a related field.
(5) Break any of these rules rather than say anything barbarous.

1.4 FURTHER READING

There have been a number of books written about or around the subject of 'community ecology'. They include Elton (1927), Allee *et al.* (1949), Dice (1952), Clarke (1954), Odum (1959), Dowdeswell (1952) and MacFadyen (1957). The concepts of 'community', 'niche', 'habitat', 'ecosystem', 'pyramid of numbers' and 'community energetics' which are fundamental to the theory of 'community ecology' are discussed in these books. Elton (1927) is a classic and is still the best short account of the subject. This book foreshadowed most of the ideas that are now current. It can be commended to anyone who is interested in the development of modern ecology. The modern idea of 'habitat' is discussed in Elton (1949) and Elton and Miller (1954). There are good discussions of the modern idea of 'community energetics' in Phillipson (1966), Clarke (1954, chapter 13), and in Odum (1959) second edition. For an introduction to statistics I would recommend Moroney, *Facts from Figures*; another good book for the beginner is Clarke, *Statistics and Experimental Design*.

2 *Environment*

2.0 INTRODUCTION

In section 1.1, I said that the scope of population ecology was best defined as the scientific study of the distribution and abundance of animals, and that distribution was merely another aspect of abundance. In this chapter I want to examine the consequences of this idea and consider how to explain the numbers that occur in natural populations. Figures 2.01 and 3.02 illustrate the numbers that were counted in two

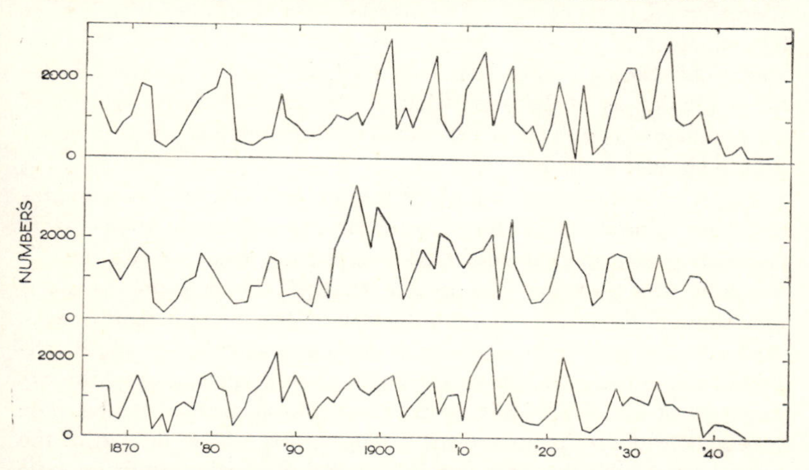

Fig. 2.01. Fluctuations in the number of red grouse shot on three moors in Scotland. (After MacKenzie, 1952.)

populations. The feature that they have in common is that the numbers fluctuate widely and erratically. This is typical of most of the natural populations that have been studied.

It is clear that if I knew what caused the population to increase and decrease in this way, and could explain the magnitude of the fluctuations, I would have resolved the problem with which I began. I shall start with the simple idea that when births exceed deaths the numbers in the population increase, and when deaths exceed births the population must decrease. Increases, decreases, births and deaths are expressed relative to the density of the population, usually either as a proportion or a per cent. If the rate of increase is called r, the birth-rate b and the death-rate d, then we may write $r = b - d$; and it follows that r is positive when b is greater than d and negative when b is less than d.

As a first step towards explaining the birth-rate and death-rate that may be observed in a natural population let us say that the birth-rate and the death-rate depend on the *environment*. This proposition is attractive because it uses familiar terms and seems to be obviously true. But closer examination shows it to contain a logical contradiction that makes it unmanageable. 'Birth-rate' and 'death-rate' are attributes of a population; so in the above context 'environment' should also be attributed to the population. Now it is well known that the birth-rate or the death-rate in a population may be influenced by the density of the population. We are trying to set up an idea of *environment* which is logically distinct from the idea of *population* because we want to be able to say that environment (through its influence on birth-rate and death-rate) determines the density of the population. So we cannot have the density of the population as part of the environment of the population: that would lead to circular argument.

To get around this problem we borrow an idea from the theory of life-insurance. The premium for a particular client depends upon his *expectation of life*, which is calculated from the age-specific death-rates in the population to which the client belongs. His expectation of life is equivalent to the estimated mean duration of life for his cohort or age-class. It is easy to see that by reversing these calculations the death-rate in a population might be estimated from a knowledge of the expectation of life of the individuals that constitute the population. The expectation of life (or to use the ecological expression instead of the actuarial one), the chance to survive, of an individual depends upon its age (i.e. it is *age-specific*); usually the very young and the very old have a smaller chance to survive than the not so young.

Similarly the birth-rate in a population may be estimated from a knowledge of the chance to multiply (or age-specific fecundity) of the individuals. Because most species have a juvenile stage when they grow without reproducing and an adult stage, when they reproduce, age-specific fecundity, or chance to multiply, is best described in relation to these two stages.

It is now possible to restate the proposition that we started with, and to say that an animal's chance to survive and multiply is determined by its environment. The chance to survive and to multiply are attributes of an individual, so the idea of environment in this context relates to an individual; there is no contradiction in thinking that an animal may have, as part of its environment, a number of individuals of its own kind that live around it. So for the purpose of analysis it is convenient to consider how the environment may influence an animal's expectation of life, speed of development and fecundity.

There is still one more qualification to be added to the definition of environment. Maelzer (1965) suggested that environment should be defined as the sum total of everything that *directly* influences the animal's chance to survive and reproduce. Maelzer pointed out that Andrewartha and Birch's (1954) definition included many things whose influence was indirect and that this idea led to ambiguity and confusion. To understand the new idea introduced by Maelzer imagine that the animal (whose chance to survive and multiply has to be explained) is caught in an *ecological web* which, by pulling one way or another, may influence the animal's speed of development, expectation of life and fecundity. The innermost strands that are joined to the animal influence it directly, and Maelzer chose to consider that these strands alone constituted the animal's environment. The outer strands, joined by diverse routes to the inner strands, influence the animal indirectly by modifying the pull of the inner strands. In Andrewartha and Birch's (1954) usage the whole web was called the environment because no distinction was made between direct and indirect influences. Whichever usage is preferred the whole web must be taken into account in studying the animal's ecology; but Maelzer's usage seems more elegant and more useful.

2.1 THE IDEA THAT ENVIRONMENT IS DIVISIBLE INTO FIVE COMPONENTS

To continue the metaphor of the preceding section. Each animal in the population that is being studied is imagined to be held in a web, the central strands of which, being joined directly to the animal, are defined as its environment. I suggest that it is convenient to recognize five such strands which I shall call the five components of environment; and I shall try to define each component so that it is homogeneous with respect to the way that it influences the animal's chance to survive and reproduce.

The five components of environment are:

resources
mates
predators, pathogens and aggressors
weather
malentities.

'Weather' and 'resources' are universal. Any animal anywhere must be experiencing stimuli which might be measured on such instruments as thermometer, hygrometer, photometer etc.; and there is no animal whose chance to survive and reproduce does not depend to some degree on its response to such stimuli. Similarly every animal, if it is to survive and reproduce, must have food, which is the characteristic and universal resource. The second component recognizes the need for every individual of a sexually reproducing dioecious species to have a member of the opposite sex in its environment to give it a chance to reproduce. This component is lacking from the environment of animals that reproduce by thelytokous parthenogenesis and of hermaphrodites. 'Predators', and 'pathogens' carry their familiar meanings, but 'aggressors' is a technical word with a strictly limited meaning. It will be discussed in chapter 5; suffice to say that 'aggressors' occur only in the environments of a few sorts of animals, mostly mammals and birds, that have evolved a highly-specialized social behaviour; aggressors, like mates, are always members of the same species. 'Malentities' is also a technical word. Malentities were first recognized as a component of environment by Browning (1962). I have departed slightly from Browning's usage, but I have retained the essence of his idea namely, (*i*) that malentities are always inimical to an animal's chance to survive and reproduce and (*ii*) that when a malentity is or is made by an animal the animal experiencing the malentity does not form part of the environment of the animal causing the malentity (chapter 7).

2.2 FURTHER READING

For other theories of environment the student should read:

Allee *et al.* (1949): physical factors, biotic factors.

Smith (1935): density-dependent factors, density-independent factors.

Varley (1947): delayed density-dependent factors.

Nicholson (1954): governing factors, legislative factors.
Clark *et al.* (1967): the life-system, conditioning effects and stabilizing mechanisms.
Morris (1963): key factors.

3 *Components of environment: resources*

3.0 INTRODUCTION

Animals, being heterotrophic, require food that has been synthesized by another organism. Microbes, plants and animals from the smallest to the largest, living, dead or decomposing, all serve as food for some sort of animal or another—such is the diversity of adaptations that have developed in relation to food. Adaptations in relation to shelter and aids to dispersal are scarcely less diverse. Shelter, even if it is no more than a place to stand on or to cling to is, like food, a universal requirement of animals. A nest in a hollow branch or a burrow under a stone are obvious shelters, but shelter may take diverse forms that may not be obvious. I have already mentioned (section 1.2) that the tsetse fly *Glossina* seems to require a 'kasenga' in which to mate; the deermouse *Peromyscus* lives along the 'edge' where chaparral meets grassland or woodland; and the butterfly *Lysandra* requires a specialized sort of place that Ford (1945) could recognize but not quite analyse. The mosquito *Anopheles culifacies* lays its eggs in rice-fields when the rice is not very high; when the rice plants reach a height of about one foot the mosquitoes disappear. While laying eggs, the mosquito weaves a tortuous path just above the surface of the water. If it is prevented by the presence of tall rice plants then it will not lay its eggs in that place. An otherwise suitable place can be made unsuitable merely by inserting glass rods that extend about a foot above the water (Macan and Worthington 1951). It seems that the selective advantage of the behaviour that leads the animal to seek out a special sort of place in which to live often depends on the association of those sorts of places with some material necessity of life, probably food or shelter; in the absence of evidence to the contrary it is convenient to count such places as resources. The requirements of animals with respect to their food and the places where they live are subtle and diverse; there seems to be no end to the subtle variations in quality, quantity and distribution that can be shown to be important. There is room barely to touch on this diversity within the scope of this book.

A good burrow for the rabbit *Oryctolagus cuniculus* should be deep to alleviate the extremes of temperature, but it should be well drained to avoid the risk of flooding during wet weather. If it is in loose sand it will need to be very deep to impede the fox which digs vertically down to the nests, but if the soil is not friable the rabbits themselves might be impeded in the underground renovations and extensions that seem to go on incessantly during the breeding season, apparently as a necessary part of the preparation for each new litter.

During the summer, the adult rabbit in non-breeding condition may thrive, even putting on weight, on its normal summer diet of straw and dry seeds, but a substantial ration of green growing herbage is necessary to keep the juveniles alive or to bring the adults into breeding condition (Stoddart and Myers, 1966). The psyllid *Cardiaspina densitexta* lives exclusively on sap sucked from the leaves of *Eucalyptus fasciculosa*; but the psyllids thrive only on a tree that is severely 'stressed', e.g. one that is seriously short of water as a consequence of a hot dry summer following a wet winter during which the tree lost most of its feeding rootlets through prolonged water-logging of the soil. At other times when the trees are not so stressed the sap is deficient in nutriment and most of the psyllids die young (White, 1966). The flea *Spilopsyllus cuniculi* sucks the blood of the rabbit *O. cuniculus*. The flea survives, but does not breed while feeding on a rabbit in non-breeding condition, because the blood of a pregnant rabbit is required to promote the development of the gonads of the flea (Mead-Briggs and Rudge, 1960). The eulophid *Melittobia chalybii* is so small that as many as 800 may find enough food to complete their development in one bee larva. The adults of *Melittobia* may be either brachypterous or macropterous. The macropterous females fly away from the place where they were reared, seeking prey into which to lay their eggs. The brachypterous females cannot fly. The first 15 or 20 eggs laid in the fresh prey emerge after about 14 days as brachypterous adults (mostly female), each of which may lay about 50 eggs into the same prey from which it has itself emerged. These eggs, together with the remaining eggs laid by the original macropterous female, give rise to larvae, all of which enter diapause which may endure for 60 days. In due course they emerge as macropterous adults which disperse. Schmieder (1933, 1939) showed that the difference between diapausing larvae giving rise to macropterous adults and non-diapausing larvae giving rise to brachypterous adults was caused by the quality of the food. The first few consume the blood and body-fluids and develop without diapause; the late-comers eat the remaining tissues and go into diapause.

These examples are striking and readily documented because they are extreme. The literature of animal husbandry abounds with information about the influence of the quality of food and shelter on the productiveness of domestic animals; the ecological literature is less advanced, but there is no need to doubt that animals in natural populations also respond sensitively to variations in the quality of their resources of food and shelter.

Large animals that disperse by walking, flying or swimming need ground, air or water to walk on, or to fly or swim through, but it would

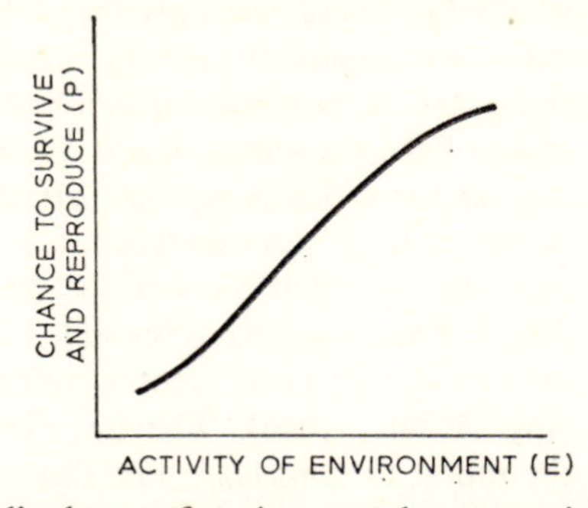

Fig. 3.01. An animal's chance of getting enough resource is enhanced when the resource is plentiful. *E* stands for the activity of the environment (resources); *P* stands for the animal's chance to survive and reproduce.

be trite to dilate on their need for these resources. Small animals that disperse by drifting or clinging (sections 10.41, 10.43) require currents that they can float in, or moving objects that they can cling to. Aids to dispersal are properly regarded as resources. The existence of many diverse and elegant adaptations for dispersal by clinging, floating or drifting is evidence of the widespread importance of these sorts of resources to the large class of animals that use them.

Because *resources* are defined as the material necessities of life which are 'used' by the animal, an animal may not be harmed by an excess of resources beyond its needs; but shortages are harmful. Figure 3.01 represents this relationship.

I mention in section 5.11 that the aphis *Eriosoma* became quite scarce in apple orchards in Western Australia after its predator *Aphelinus* had been introduced to that area. The few *Eriosoma* that now live in the orchards are surrounded by easily accessible food in great abundance. This is the usual condition of most animals: some component of environment other than food presses on them so heavily that they are unlikely to become numerous enough to press heavily on their stocks of food (section 4.2). The same may be true of other resources besides food.

The converse also happens. Certain species seem to have the capacity to multiply right up to the limit of their supplies of food (or other resource) at least locally. This resource then becomes an important component in their environment (sections 3.12, 3.2).

A third possibility is that a resource may be present yet inaccessible, perhaps because the animals may not be good at finding it. In these circumstances many animals in the population may die for lack of the resource despite the fact that some, perhaps a large proportion, of the resource remains unused. Andrewartha and Birch referred to this paradox of scarcity amidst plenty as *relative shortage* (section 3.1) to distinguish it from the *absolute shortage* which arises when the resource is fully accessible but insufficient in quantity (section 3.2).

3.1 RELATIVE SHORTAGES

3.11 *Extrinsic relative shortages*

Jackson (1937) was probably the first ecologist to draw attention to the circumstances in which most of the individuals in a population may go seriously short of food even though they are surrounded by more than enough to sustain the whole population. The tsetse fly *Glossina morsitans* sucks blood, chiefly from the larger species of antelopes. It requires to feed frequently, especially during hot weather, but it does not stay near its food after engorging. So each meal is preceded by an independent search. If antelopes are sparsely distributed, a fly may have little chance to find sufficient food to produce many or any offspring before it dies from starvation, despite the fact that the blood in one antelope would be enough for many flies. Jackson pointed out that there would be a particular density in the population of antelopes that would tend to maintain births equal to deaths in the population of *Glossina*. Changes in the numbers of antelopes would occur quite independently of the number of *Glossina* feeding on them but would nevertheless determine the rate of increase or decrease in the population of flies. Andrewartha and Browning called the sort of shortage experienced by *Glossina* an *extrinsic relative shortage* to distinguish it from the sort of shortage (discussed in section 3.12) in which the shortage is caused by the animals' pressing heavily on the resource.

Potts and Jackson (1953) brilliantly confirmed Jackson's hypothesis when they exterminated three species of *Glossina* in 600 acres at Shinyanga by shooting most but not all the 'game' in the area. About 75 hunters were employed to shoot as many as possible of the larger mammals which might serve as hosts for tsetse flies. At the beginning of the experiment there were about 10 such animals to the square mile. During the next five years about 8,000 animals were shot, including many that came in from outside the area. Towards the end of this period the numbers of the more important species had been greatly reduced, but there was still a small herd of elephants and there were many of the smaller species of ungulates still living in the area. Altogether the animals still living there at the end contained many times more than enough blood to support all the tsetse flies that had been living there at the beginning. Nevertheless the animals were few enough and distributed sparsely enough to give the tsetse flies only a small chance of finding enough food. By the end of the exercise the authors were satisfied that *G. morsitans* and *G. swynnertoni had* been exterminated and *G. pallidipes* had either been exterminated or reduced to extremely low numbers.

The amount of food in the area was independent of the number of flies because those few that did find food did not, by virtue of their feeding, make any difference to the amount or distribution of the food. Also, the flies' chance to get enough food to survive and reproduce was independent of the size of the population of flies. There was no shortage of food in the absolute sense but the relative shortage of food was so severe that the population became extinct.

The ticks *Ixodes ricinus* that live in the rough upland pastures of northern England feed mostly on the blood of sheep and only to an unimportant extent on that of wild animals (Milne, 1949, 1951). During most of the year the ticks remain inactive in the mat of vegetation near the ground. During spring they emerge from this shelter and climb up an exposed grass-stem. If, during the 9 or 10 days that they can survive in this exposed position, they are picked up by a sheep as it brushes by, they cling to the sheep until they are engorged with blood. They then drop from the sheep and crawl into the mat of vegetation again. They digest their meal and develop to the next stage of the life-cycle. Each tick requires to feed three times during its life – as a larva, as a nymph and as an adult. After the third meal the eggs are laid.

On one farm on which Milne worked he estimated that a tick had about 40% chance of being picked up by a sheep in any one year. The probability that a tick would be picked up three times during its life can be estimated as $0{\cdot}4^3 = 0{\cdot}06$. In other words a tick might have a 6% chance of getting enough food to allow it to complete its life-cycle and lay eggs. This farm carried about one sheep to the acre. Had there been more sheep the ticks would have had a greater chance of getting enough food because a tick's chance of being picked up depended only on the number of sheep in the area and their activity. It did not depend at all on the number of ticks in the area because the feeding of the few that did get a meal made no difference to the number of sheep in the area. In other words there was a severe shortage of food in the relative sense but no shortage whatever in the absolute sense.

In section 1.1, I mentioned that in south-eastern Australia the distribution and abundance of *Thrips imaginis* can be very largely explained in terms of weather. To a certain extent weather influences the thrips' chance to survive and multiply directly because dry weather during summer increases the likelihood that the pupal stage will die from desiccation, and cool weather during winter reduces fecundity and retards development. But the weather is chiefly important for its indirect influence on the supply of food for the thrips. In order to explain how this happens it is necessary to describe the life-cycle of *T. imaginis*.

The nymphs and adults live almost exclusively in flowers; they suck the sap from stamens, pistil and petals but in particular they suck the contents of the pollen grains. Pollen is essential; without it they do not grow or lay eggs. The numbers fluctuate regularly with the seasons: they reach a maximum about the end of November (southern spring) and are usually at a minimum during February–March (late summer) and during July–August (winter). The fluctuations are large: in November I have counted as many as 2,000 in a sample of 20 small roses, but in July or August it is usual to find about 50 in a similar sample.

The eggs are laid in the tissues of the flower. At 20°C the eggs hatch in about four days. The nymphs feed on the tissues of the flower for about six days; then they crawl into some sheltered place usually among the debris on the ground and pass through two non-feeding 'pupal' stages which take about five days. At lower temperatures all stages take longer. The newly emerged adults fly in search of a place to live and lay their eggs. Their chance of finding one depends on the number and distribution of flowers in the area.

The number of flowers in the area fluctuates regularly with the seasons. In southern Australia the weather resembles that of the Mediterranean region of Europe. During the long, dry summer, annuals wither and perennials produce few flowers. The winter is the time for growth, especially of annuals, but still there are few flowers. But in the spring there is a great profusion of flowers not only on the numerous and widespread annuals but on many perennials as well. A thrips, taking wing at this time of the year, could scarcely fail to find a suitable flower in which to live for a while and lay its eggs. The survival-rate and the birth-rate are high and the numbers increase rapidly. But the flowers increase even more rapidly so that, even with the expanding number of thrips, there is little chance of an absolute shortage of food except perhaps locally for a brief period.

But the period of great abundance is short-lived. The change that comes with the summer is dramatic. Food becomes scarce and remains so, no matter how little is eaten, because the weather has changed. Flowers that might provide suitable food and breeding places for *Thrips imaginis* are few and widely scattered. A thrips that takes to the wing in these circumstances has little chance of finding food. Many of them die without reproducing. There is still no absolute shortage of food because our samples at this time of the year showed only two or three thrips per rose; and any other flower that was examined showed thrips present at similar low densities. The population which was large in November declines rapidly as summer advances and continues, with perhaps a minor respite in the autumn, to decline through winter until the trend is reversed with the onset of spring (Fig. 3.02).

The great seasonal changes in the abundance of the thrips are related to the seasonal changes in the accessibility of their food. At no time is there an absolute shortage of food, but at certain seasons of the year food becomes inaccessible (i.e. unlikely to be found) merely by becoming sparsely distributed (Davidson and Andrewartha, 1948a, b).

Criddle (1930) found that in Manitoba the numbers of grouse increased during outbreaks of grasshoppers, chiefly because the grasshoppers provided abundant food for the grouse when they were nesting. Outbreaks of grasshoppers tend to develop during warm, dry weather, and often come to an end after a spell of cool, damp weather. But the numbers of grouse declined during prolonged cool weather even before the numbers of grasshoppers had changed very much. This was because the grasshoppers were less active during cool weather and the grouse had greater difficulty in finding or catching grasshoppers when they were inactive. In these circumstances many nestlings died from lack of food. There was a relative shortage of food that had been caused, in the first place, by the influence of weather on the behaviour of the grasshoppers.

A relative shortage of food may have been an important component in the environments of primitive men. The following quotation from Braidwood and Reed (1957) makes this point.

> The main point here is that, as Deevey (1956) has already shown, the population of the primary hunting and collecting culture is not limited by the potential amount of food, but by man's ability to get it during the

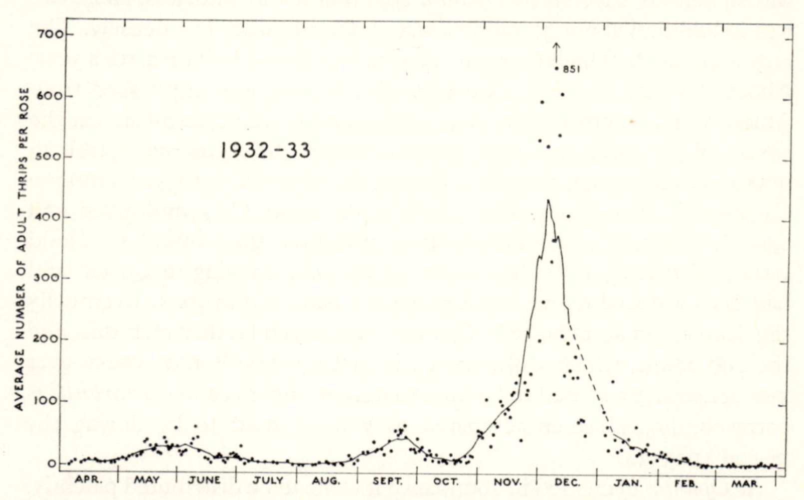

Fig. 3.02. The annual cycle in a natural population of *Thrips imaginis*. The points represent daily records. The curve is a 15-point moving average. (After Davidson and Andrewartha, 1948a.)

> leanest season. Proto-Magelmosian man was probably often hungry, even starving, in country which should have indefinitely maintained a population at least 100 times (10,000%) as high (or 38 people to the square mile). That no hunting and gathering culture ever approached this population optimum . . . is an indication of the human technological level of the hunting-collecting level.

It is characteristic of extrinsic relative shortages that the shortage is caused by what is obviously an outer strand of the ecological web. With the grouse, temperature in the environment of the grasshoppers, with *Thrips*, weather in the environment of the food-plants, with *Ixodes* probably food in the environment of the sheep and so on.

Intrinsic relative shortages are best explained in the same way but the explanation is more subtle (sections 3.3, 9.221).

3.12 *Intrinsic relative shortages*

Prickly pear is the name given to several species of *Opuntia* that were serious weeds in north-eastern Australia. They were originally brought from America and planted in gardens. They escaped from the gardens and spread into agricultural and pastoral land. Dodd (1936) reported that in 1925 some 30 million acres of potentially good agricultural land was so densely covered by *Opuntia* spp. that it was valueless, and there was in addition about 30 million acres infested rather less densely. The area occupied by *Opuntia* was increasing by about a million acres a year. About this time the moth *Cactoblastis cactorum* was introduced from America. The moth lays its eggs on the cactus; the caterpillars eat the tissues of the cactus and their presence makes the cactus susceptible to micro-organisms that cause rot. During the next five years 3,000 million *Cactoblastis* were bred and set free in the area. They multiplied and spread, destroying the prickly pear wherever they found it. Dodd estimated that by 1934 the amount of *Opuntia* growing in Queensland had been reduced to one tenth of what it had been in 1925. Eventually the decrease in abundance of *Opuntia* went much further than this, and the end result, which still persists, is that *Opuntia* is now scarce over vast areas where it used to be inordinately abundant; and *Cactoblastis* is correspondingly scarce compared to what it used to be during the period 1925–34.

As *Opuntia* decreased in abundance it came to be distributed patchily. Because *Opuntia* is so patchily distributed, some of it goes undiscovered while many *Cactoblastis* die without finding food for their young. So the shortage is a relative shortage. Because the scarcity of *Opuntia* and its patchy distribution was caused in the first instance by the feeding of *Cactoblastis* we call the shortage an *intrinsic relative shortage*.

According to Smith (1939) extreme patchiness is characteristic of the distribution of any weed or insect pest that is under biological control, i.e. it is being kept scarce by the feeding of an active 'predator'. And Flanders (1947) pointed out that the characteristic patchiness in the distribution of the 'prey' was a consequence of two qualities that are characteristic of all successful agents of biological control: (*a*) They can outbreed, and therefore usually destroy, their 'prey' in local situations. (*b*) They are highly dispersive relative to their 'prey' and are good at searching.

Cockerell (1934) described the circumstances in which certain species of scale insects were kept in low numbers by predators.

I recall some observations on Coccidae made in New Mexico many years ago. Certain species occur on the mesquite and other shrubs which exist in great abundance over many thousands of square miles of country. Yet the coccids are only found in isolated patches here and there. They are destroyed by their natural enemies, but the young larvae can be blown by the winds, or carried on the feet of birds, and so start new colonies which flourish until discovered by predators and parasites. This game of hide-and-seek doubtless results in frequent local extermination, but the species are sufficiently widespread to survive in parts of their range, and so continue indefinitely.

In section 5.11 I describe how the predators *Rodolia* and *Ptychomyia* press heavily on their food *Icerya* and *Levuana*. The relationship between these predators and their food differs in details, one from the other and each from *Cactoblastis*, but the general pattern of the relationship is the same in each instance. The food is distributed patchily in small temporary colonies. The colonies usually have a short life because the highly dispersive predators are likely to find them soon after they have been started, and having found them, are likely quickly to destroy them. In each local situation where the predators find a colony of the prey the predators at first increase in numbers and later decrease towards or perhaps to zero under the influence of an absolute shortage of food. Other species that are not so good at dispersing and searching, yet are good at destroying the local colonies once they have been found (e.g. *Rhizobius*, section 5.11), may experience an equally severe relative shortage of food without depressing so severely the abundance of their food (Andrewartha and Browning, 1961).

3.121 Outbreaks of pests

Many of the common insect pests of our farms and gardens have the same potentialities as *Cactoblastis*. Indeed the story of the vine aphid *Phylloxera vastatrix* is like that of *Cactoblastis* except that it has a different ending. This aphid was accidentally introduced into France from America through the port of Bordeaux about 1865. Like *Cactoblastis* it spread quite rapidly and its spread was assisted by man (though inadvertently in this instance); and it killed out whole populations of its host. It destroyed 3 million acres of vines in France before someone discovered that it could be checked by grafting European vines on American root-stocks.

The woolly aphis *Eriosoma lanigerum* has a similar potential with respect to its host the apple; but it is nevertheless quite scarce wherever apples are grown. Apples are now universally grafted on resistant root-stocks; so *Eriosoma* does not thrive on the roots; above ground it is heavily preyed on by *Aphelinus* (section 5.11). From time to time outbreaks of other pests such as locusts, thrips, scale insects and army worms, rodents and rabbits remind us that they too, in the absence of other checks, would be able to multiply until they were checked by an absolute shortage of food.

It is not often that one can see a population of codlin moth *Cydia pomonella* that is checked by an absolute shortage of food because usually the population is held at a much lower level by repeated poisoning. But in 1952 I came across an orchard in which the codlin had become highly resistant to DDT. Although the orchard had been sprayed 10 times during the summer this was virtually equivalent to not spraying it at all. The codlin multiplied until, towards the end of the season, virtually every apple contained a caterpillar and a few contained more than one. Because the seeds of the apple are an essential part of the diet and because only the seeds of a large apple contain enough food for more than one caterpillar there can be little doubt that virtually all the food had been eaten. Most of the apples showed signs of the burrowings of at least several caterpillars indicating that many had died from starvation. In the local situation represented by this 10 acres of orchard, all the food was accessible but there had not been enough to go round. The shortage was *absolute* and severe. But for reasons that will be discussed in sections 3.21 and 3.22 the consequences for both the apple-trees and the codlin moth were different from those described for *Opuntia* and *Cactoblastis*.

3.2 ABSOLUTE SHORTAGES

There are three statements which might be made about an absolute shortage of food (or other resource), two of which seem to be relevant to ecology. (*a*) The statement that the animals experience an *absolute shortage* of food (or other resource) emphasizes the fact that some of the animals fail to get enough of the resource because there is not enough to go round. This statement focuses attention on the animals in the present generation, their chance to survive and reproduce. It might be an important comment about any of the species mentioned in section 3.12. (*b*) The statement that the animals have become so numerous that *a large proportion of the resource is being consumed* focuses attention on the future of the resource. For example, the consumption of a large proportion of *Opuntia* by *Cactoblastis* or of *Icerya* by *Rodolia* in one generation may impair the regeneration of the resource and so result in an even greater shortage of food for the next generation. This statement is scarcely relevant to the population of codlin moth discussed in section 3.12 because next summer there will be another crop of apples independent of how many may have been eaten by the present generation of codlin moth, i.e. the regeneration of the resource is not impaired by heavy consumption of it. (*c*) The statement that *intraspecific competition* is taking place emphasizes the fact that some of the animals are better than others at getting enough of a resource that is in short supply. This statement is relevant to ecology only indirectly in so far as it relates to the evolution of the animals that the ecologist might wish to study. This consequence of crowding will not be pursued. The other two consequences will be analysed a little further.

Andrewartha and Browning (1961) described four case-histories which show how different sorts of resources may vary in their response to use by a dense population and how animals may vary in their response to an absolute shortage of resource. I quote these four case-histories in full.

(*i*) The blowfly *Lucilia sericata* may be reared in the laboratory on meat. If the number of maggots on a given quantity of meat is carefully regulated (so that there are enough to eat all the meat, yet not so many that there is too little meat to go round) it is possible to rear about 1,600 well grown maggots from one kg of meat. However if the density of the population is increased a little there will be an absolute shortage of food, yet some maggots may still be able to complete their life-cycle. If the density is increased still further until the absolute shortage becomes extreme, all the food may be consumed without any maggot getting enough to complete its life-cycle (Ullyett, 1950). In these circumstances all the food is 'wasted' because none is used to produce parents for another generation.

In nature, the number of maggots that are found in a carcase often exceeds the optimum. Waterhouse (1947) exposed the carcases of 27 sheep near Canberra and allowed the wild flies to lay eggs on them freely. Many more eggs were laid than the carcases could support and the absolute shortage of food was extreme. The carcases produced from 109 to 67,000 adult blowflies with an average of 10,000. At the optimum rate mentioned above a carcase weighing 50 kg might have produced 80,000 adult flies. The difference may be attributed to the severe absolute shortage of food. A number of species were using the carcase at the same time and this complicates the full explanation of the results (section 7.2), but there can be no doubt of the severe absolute shortage of 'effective' food (see footnote to Table 3.01) for that generation of maggots, and its influence on the size of the next generation.

The supply of food for the next generation depends on the number and size of carcases in the area. This is determined by the number of deaths among sheep and other suitable animals in the area which is, of course, quite independent of the density of the population of maggots in the carcases that bred the last generation. In this respect blowflies are typical of a large number of species that eat carrion, refuse, or 'litter' of any sort, or that use such material for building nests. For such species the supply of resource for future generations is not influenced by the density of the population of animals using the resource in the present generation; but the size of the next generation depends upon the absolute shortage experienced by the present generation (see section B(*a*) of Table 3.01).

(*ii*) Leopold (1943) found that the numbers of deer on the Kaibab plateau in Arizona were determined largely by predators until most of the predators were shot. At first the deer increased greatly in numbers and almost ate themselves out of food. At the same time much of the herbage that had been grazed and trampled did not regenerate. The deer became fewer but did not die out because there was always *some* food to be found. Their numbers came to be determined largely by the distribution and abundance of their food and especially its accessibility. In this case the numbers of animals in the present generation, by virtue of their feeding, determined the distribution and abundance of the food for future generations; and the absolute shortage of food experienced by the present generation reduced the number of young born into the next generation. That is, the deer responded to an absolute shortage of food in the same way as the blowflies; but their food responded to the number of animals using it, unlike the carrion which was food for the blowflies (see section A(*a*) of Table 3.01).

(*iii*) The bug *Nezara viridula* used to be a serious pest of beans, tomatoes, and other crops in parts of South Australia until the egg-parasite *Asolcus basalis* was introduced and became established. The females of *Asolcus* rarely oviposit into an egg of *Nezara* that already contains an egg or larva of *Asolcus*. Even if they do, only one larva will survive because the other gets eaten along with the contents of the egg. This behaviour ensures that no matter how great the absolute shortage of eggs of *Nezara* may be, relative to the numbers of *Asolcus*, nevertheless none will be wasted on individuals that do not get enough food to complete their life-cycle. So the numbers of individuals of *Asolcus* in the next generation does not depend on the absolute shortage experienced by their parents but merely on the amount of food that was present. In this respect

Asolcus differs from both blowflies and deer. The food of *Asolcus* resembles that of the deer (but differs from that of the blowflies) in that the supply of food for future generations depends on the numbers of *Asolcus* seeking food in the present generation.

(*iv*) According to Kluijver (1951) a population of tits *Parus major* living in a mixed woodland of young trees were few relative to another population living nearby in a woodland of old trees. The young trees had few holes suitable for nests. Nearly all the holes were used, so, despite their low numbers, the tits must be considered to be experiencing an absolute shortage of nesting-sites. The territorial behaviour of the birds ensured that a pair that managed to obtain a nest had the unrestricted use of it during the breeding season. Thus each nesting-site that was used at all was used to rear potential parents for the next generation. In this way the tits' response to an absolute shortage of a resource resembles that of *Asolcus* and differs from that of both deer and blowflies. Because

Table 3.01. *Classification of the relationships that occur between animals and their resources in conditions of absolute shortage (After Andrewartha and Browning, 1961)*

A. The supply of resource for future generations depends on the number of animals in the present generation using the resource

(*a*) The amount of resource used effectively* by the present generation depends upon the absolute shortage that they experience.	(*b*) The amount of resource used effectively by the present generation is independent of the absolute shortage that they experience.
Deer on Kaibab, *Cactoblastis*, cats on Berlenga.	*Asolcus*.

* 'Effective' is used in the sense defined by Andrewartha and Birch (1954, p. 498), i.e. it refers to that part of the resource that is acually used by those individuals that get enough to mature and so have a chance to contribute to the next generation.

B. The supply of resource for future generations is independent of the number of animals in the present generation using the resource

(*a*) The amount of resource used effectively by the present generation depends upon the absolute shortage that they experience.	(*b*) The amount of resource used effectively by the present generation is independent of the absolute shortage that they experience.
Lucilia, certain parasites, most 'detritus feeders'.	*Parus*, codlin moth, certain parasites.

the nesting-sites are used without being destroyed the supply of nesting-sites for the next generation is independent of the number of tits that sought to use them in this generation. Nesting-sites for tits thus resemble food for blowflies and differ from food for both deer and *Asolcus*.

The relationship between the codlin moth and its food is similar to that between *Parus* and its nesting-sites because (*a*) its presence in an apple does not influence the size of the crop in the following year and (*b*) its behaviour ensures that none of the essential food (the seeds of the apple) is wasted on any individual that does not get enough to grow to maturity (section 3.211).

Andrewartha and Browning classified these four case-histories into the four classes in Table 3.01 and suggested that this classification may apply quite generally to all natural populations that become abundant relative to any sort of material resource.

Animals that come into category A of Table 3.01, by virtue of their relationship to their resource, can maintain the condition of absolute shortage only locally and temporarily; over any large area the shortage develops into an intrinsic relative shortage (section 3.12). On the other hand, animals that come into category B may, if they are not exposed to an extrinsic relative shortage of a resource nor limited by any other component of environment, maintain the condition of absolute shortage over large and widespread areas. Animals that fall into categories A(*b*) and B(*b*) of Table 3.01 are adapted to make the best use of a resource that is likely to be absolutely short. These adaptations, though extremely diverse, are best classified together under the general heading *territorial behaviour*.

3.21 *Territorial behaviour in relation to resources*

An obvious feature of territorial behaviour is that an animal defends a 'territory' against another individual of its own species thereby enhancing its own chance of getting enough resource at the expense of the other, which would seem simple enough. But the ecological consequences of territorial behaviour and the evolutionary explanation of it are not simple or well understood. The contests are often not lethal or even physically damaging, a song, a patch of colour or an odour serving instead of biting, scratching or pushing. The outcome of the contest is often a foregone conclusion determined by an accident of age, experience or residence which seems not to be closely, if at all, related to the heredity of the contestants. There are often obvious advantages accruing to the winner; but the loser often seems to be so willing to lose that it would seem that there must be some balancing selective advantage in being a loser too. Finally it is necessary to explain how territorial behaviour on the part of individuals leads to the whole population using a resource more economically than it might in the absence of territorial behaviour. Is this merely a by-product of the conventional mechanisms of natural selection? or is there some mysterious mechanism of 'group selection' operating as well? It is beyond the scope of this book to explore these interesting questions. The several examples discussed below merely indicate a few of the complexities that might have to be taken into account.

3.211 Territorial behaviour in insects

The eggs of the bug *Nezara viridula* are laid in a 'raft' of up to 90 barrel-shaped eggs standing upright and side-by-side. A single egg contains enough food to allow a larva of *Asolcus basalis* to grow to maturity. The female *Asolcus* pierces the egg of the bug with her ovipositor and lays her own egg inside the egg of the bug. If more than one egg is laid into the same egg of *Nezara* one of the larval *Asolcus* will eat the others so that usually one, but never more than one, *Asolcus* emerges from a *Nezara* egg that has had one or several *Asolcus* eggs laid into it. But according to Wilson (1961) it is unusual for a *Nezara* egg to have more than one *Asolcus* egg laid into it.

Wilson described how a female *Asolcus*, ovipositing on a raft of bug eggs, will examine each egg thoroughly with her antennae before either inserting or not inserting her ovipositor into the egg. If she pierces the egg she usually (on 93% of occasions in one experiment) lays an egg into it. If she lays an egg she invariably marks the bug egg with a pheromone, 'painting' the pheromone over the top of the bug egg by stroking the tip of her ovipositor over it. The presence of this mark is highly effective in inhibiting herself or any other female from laying another egg into a marked bug egg. Table 3.02 gives the distribution of 'stings' when 75 eggs of *Nezara* were exposed to either one or two females of *Asolcus*. A 'sting' was counted when a female was seen to pierce an egg with her ovipositor. On five occasions out of 85 a female pierced an egg without laying an egg into it. The values of χ^2 at the foot of the table show no significant difference between the two empirical frequency distributions; but both of the empirical distributions were significantly different ($P \ll 0.001$) from the theoretical Poisson distribution.

When more than one female was present on an egg-raft they took little notice of each other at first but towards the end when most of the bug eggs had been marked one female usually became highly aggressive, driving all the others away; she then continued searching on her own for unmarked eggs.

The important feature of this behaviour is not so much the marking of the egg with pheromone by the fortunate first-comer but rather the capacity of the late-comers to recognize the mark and so avoid wasting their eggs. The eggs laid after the first one will be eaten; so the first-comer is assured of a representative in the next generation without making the mark; but the latecomers have no chance of contributing to

Table 3.02. *Distribution of 'stings' in 75 eggs of Nezara exposed to 1 or 2 females of Asolcus (After Wilson, 1961)*

No. of stings (x)	Number of eggs (f)			
	Observed		Expected on a random distribution	
	1 female	2 females	1 female	2 females
0	1	2	23	21
1	66	57	27	27
2	5	11	16	17
3	1	4	6	7
4	2	1	2	3
χ^2	Non-significant		86·5	55·4

the next generation by laying their eggs in a place that is already occupied, whereas they may have a chance of finding an unoccupied one if they continue the search elsewhere. Of course the first-comer, by marking the egg, enhances its own chance of having more than one representative in the next generation; but, after the first egg has been laid, the mark merely puts the first-comer on the same level as the late-comers.

The males of *Asolcus* are also territorial. The life-cycle of *Asolcus* from egg to adult takes about 11 days at 27°C, but the males usually emerge about one day before the females. The first male to emerge takes possession of the egg-raft, driving all others away. If he should leave the egg-raft vacant (which may happen if there is a prolonged period with no female emerging) any other male arriving on the vacant egg-raft will take possession driving all others away, including the original occupant if he should happen to return. The conflicts though vigorous never cause physical hurt to either combatant. The male that is in occupation copulates with each female as she emerges. The territorial behaviour of male *Asolcus* is difficult to explain without postulating that the loser has more chance of contributing to the next generation by running away to seek a vacant egg-raft than by staying to fight with a male already confident by virtue of the ownership of a territory. The sex ratio in *Asolcus* is 1 male to 5 females, which may make the hypothesis more plausible.

The adult codlin moth *Cydia pomonella* lays its eggs on the outside of an apple. The caterpillar, on hatching, burrows into the core of the apple where it remains, feeding largely on seeds, until it is mature. Then it cuts a way out of the apple and seeks a place to pupate. Usually *Cydia* is rare relative to its food because, in a well-kept orchard, it runs a big risk of being poisoned by insecticide; and in a neglected orchard, it is likely to experience an extrinsic relative shortage of food. Consequently an apple rarely contains more than one caterpillar, despite occasional signs of more than one entry into the same apple. But artificially, by withholding insecticide from an otherwise well-run orchard it is possible to provide *Cydia* with an abundance of readily accessible non-poisonous food. In these circumstances *Cydia* may multiply beyond the limits of its food, and many caterpillars may seek to feed in one apple at the same time. Then the territorial behaviour of the caterpillars becomes clearly manifest; the core is divided into 'territories' which contain just enough food for a larva to grow to maturity. The unsuccessful contestants die; most of them are probably eaten by the successful ones. Geier (1963) found that the core of a 'small' apple would support

Table 3.03. *Reduction in size of territory in response to crowding in* Cydia pomonella *(After Geier, 1963)*

Initial no. caterpillars per apple	Small apples		Large apples		
	Mean no. mature larvae per apple	Mean wt. adult females (g)	Mean no. mature larvae per apple	Mean wt. adult females (g) diapausing	Mean wt. adult females (g) non-diapausing
1	0·93	26·9	0·93	32·7*	22·2
2	1·11	26·1	1·54		
3 or more	1·13		—	—	—
3–5			2·25	29·8*	21·6
6–10			2·26		

* Difference significant $P < 0{\cdot}01$.

only one or occasionally two caterpillars, but the core of a 'large' apple would support two or three (Table 3.03). Apparently the caterpillars showed little tendency to defend a territory larger than their needs, but even under intense crowding they could rarely be forced to accept a territory that was too small for their needs. Only the crowded diapausing larvae developed into adults that were significantly smaller than the uncrowded ones.

3.212 Territorial behaviour in vertebrates

The Australian magpie *Gymnorhina dorsalis* is a black and white bird about the size of a pigeon. It is one of the commonest birds in Australia. Carrick (personal communication) studied the ecology of a population of magpies in several thousand acres of farmland near Canberra. Carrick found that the birds in this area comprised two groups which he called the 'tribes' and the 'flock'. A tribe was a group of 2 to 12 birds which lived in and defended a 'territory'. The boundaries of the territories changed a little during the years as a result of vicissitudes that happened either to the tribe itself or to its neighbours. But, in general, a tribe, once it had established itself in a territory, remained there for the duration of its existence as a tribe. The territory sufficed for all the requirements of the tribe. They found all their food within its boundaries and reared their young in it. The minimum requirement for a territory was a tree in which to build a nest, but most territories contained many trees. Carrick rarely found a bird from the tribes that was undernourished, so apparently each territory contained abundant food for the occupants. Nevertheless, the size of the territory bore little relationship to the size of the tribe or to anything else that could be measured. It seemed as if the size of the territory depended chiefly on the innate qualities of the birds, especially their capacity to 'dominate' their neighbours. An established tribe rarely accepted an immigrant except occasionally to replace a female that had died, nor allowed one of its own offspring to remain in the territory after they had matured. Probably the tribes never lost any member by emigration, but this was more difficult to measure. Individuals died, but the survivors continued in the territory until such time as they could no longer defend it; then the whole group would disappear. Sometimes the death of a single bird would lead to the summary eviction of the survivors, perhaps because the one that died had been the leader. Occasionally Carrick found birds from a tribe that had been 'evicted' living as individuals in the 'flock'.

The other group of birds, which Carrick called the 'flock', comprised all those that were not members of a tribe with an established territory. None of these birds bred. Apparently the stimulus of the changing seasons was in itself not sufficient to bring the birds to full sexual maturity; they needed, in addition, some stimulus that came from being a member of a tribe in an established territory. The numbers of the flock were swelled each year, towards the end of winter, by the influx of the young birds that had been reared by the tribes during the previous spring. The flock was also increased from time to time by the remnants of tribes that had abandoned, or been driven out of their territories. From time to time a 'tribe' would originate in the flock and establish itself in a territory.

The birds in the flock were more likely to die from parasites and diseases than those in the tribes, and their supply of food was less certain, but the flock persisted because of the regular influx of young birds from the territories each year.

Kluijver (1951) described 'territorial' behaviour in a population of great tits *Parus major* living in a mixed deciduous wood near Arnhem, Netherlands. At the beginning of autumn the population comprised old birds that had nested there during summer and their progeny. At this time of the year the birds formed pairs and each pair defended a territory. The 'fighting' was not overt. A bird would fly towards its opponent and turn upwards in such a way as to display a patch of colour on its throat. In the end the 'weaker' bird would give up and fly away. Usually a substantial proportion of the population would be driven out in this way. Once they had been driven out of the country that was familiar to them they became more vulnerable to predators, exposure and starvation. At the beginning of winter there were left in the wood a number of pairs, each with its own territory. A bird defending a territory that it had owned the previous year seemed to have the advantage over a stranger or an upstart. So, many territories were held by the same birds for more than one year. The minimum requirement for a territory was a hole in a tree or some other suitable place for making a nest. But when Kluijver artificially increased the number of places for nests by putting nest-boxes in the wood, the number of territories did not increase up to the limit set by the number of boxes. Nor did they depend directly on the amount of food because at the time when they were being established there was a lot more food present than there would be during winter; yet there were rarely enough birds left in the wood to exhaust the supplies of food during winter. The size and number of territories seemed to depend on something else that could not be measured. Like the

Australian magpies the great tits seemed to possess innate qualities of 'aggressiveness' or 'unsociability' which largely determined the size of the territories and hence the number of birds left to overwinter in the wood.

During winter the pairs disbanded and the birds disregarded territories. Each bird ranged over a much wider area seeking food and, in a place where food was abundant, a number would congregate without any sign of 'unsociableness'. But each bird would recognize without defending it, a smaller, usually a central part of the country that it ranged over in which it would find places for roosting and sleeping, and in which, in due course, it would choose a nesting site. Kluijver called this area its 'domicile'. The domiciles quite often overlapped and were usually larger than territories which never overlapped.

With the approach of summer, pairs were re-formed and territories were resumed briefly. When the nestlings had to be fed the boundaries of the territories were largely ignored and the birds ranged widely in search of food.

The tits were like the Australian magpies, in that the size of the breeding population was determined by the 'territorial' behaviour of the birds; that is, the territories seemed not to be readily compressible under the influence of crowding. They differed from the magpies in that: (*a*) the territories were held by pairs instead of tribes, (*b*) the tits defended territories only briefly, ignoring the boundaries and ranging widely over what had been the territories of others when it came to seeking food for nestlings during summer or themselves during winter.

Territorial behaviour in the European rabbit in Australia has been studied by Mykytowycz (1958, 1959, 1960, 1961, 1968) and by Myers and Poole (1959, 1961, 1962, 1963a, 1963b). The rabbits breed during the winter when they can feed on green growing herbage. At the beginning of the breeding season females in breeding condition begin to renovate and extend burrows and to undertake other home-making activities. In this condition females are both gregarious and aggressive towards each other. The presence of a female digging, especially in an old, well-developed warren, attracts other females. They fight fiercely and some may be driven away, but some persist. The end result is that a group of 3–5 females is formed which claims joint possession of a cluster of burrows subject to the proviso that one female dominates all the others, taking first choice of the best sites in the territory; the others also arrange themselves in a hierarchy of dominance. It is an uneasy but persistent truce. Meanwhile males fight fiercely among themselves for 'ownership' of a group of females. As the fighting subsides one emerges as the victor; most of the contestants have left; but usually one or two remain; they seem to be accepted by the victor because they accept subordinate rank with scarcely any attempt at retaliation. If there is more than one subordinate one will dominate the others. So the typical social group comprises 2–3 males and 3–5 females, the members of each sex being arranged in a social hierarchy clearly dominated by the individual with the highest social status.

The outcome of the contests for status depend largely on tradition, age and weight. When the contest takes place in an established territory it is mostly the yearlings from the previous breeding season that are driven out; and usually last year's dominants retain their status. When the contest occurs in new territory victory usually goes to the oldest or the heaviest. (Usually age and weight go together.) It has not been shown that breeding has much influence on an animal's chance of becoming a dominant. But it has been observed that at least some of those that are slow to find a place in a social group are so because they aspire to the dominant position but have not been successful in the first few groups that they tackled. On the other hand some of those that fail to enter a group seem to belong to the other extreme.

Each group defends a territory centred on its cluster of burrows. The territory is marked with pheromones. A secretion from the inguinal gland is smeared on faeces which are dropped in conspicuous and strategic places especially on the main pathways leading from the burrows. Sticks, stones and other landmarks, even heaps of old faeces, or any faeces dropped by a stranger, are marked with the secretion of the chin gland; and urine is also used. The dominant male is most active in marking, so his odours pervade the territory.

The territory, once marked, is defended strongly, especially during the breeding season. All members of the group will join in the defence of the territory, but individuals react more strongly against intruders of their own sex. The dominants of both sexes are most vigorous in defence. The burrows and the immediate approaches to the burrows are policed most vigorously; as might be expected the strictness and the vigour of the defence declines with distance from the burrow. It is likely that the characteristic odour of the group and its territory gives confidence to the members of the group and conversely saps the confidence of the intruder. Certainly this impression is confirmed by watching the behaviour of an intruder whose demeanour changes to one of marked and obvious wariness and diffidence as soon as it can be seen to pick up the odours of a strange well-marked territory. Because the odours are stronger near the burrows the defenders are fiercer and more persistent the closer they are to the burrows and the intruder correspondingly less so; but as the fight or the chase approaches the boundary of the marked territory the difference between the antagonists might disappear or might be reversed.

Not only are the individuals in a social group bound to their territory by the odours of the dominant male but they are also bound to each other in the same way. As a pregnant female approaches her term, the dominant male consorts with her driving away all other males; during the post-partum oestrus he copulates with her. During courtship and after copulation he urinates over her and smears the secretion from his chin gland on her. At other times he 'chins' the females, the kittens and the juveniles in the group. It seems that by chinning the kittens and the juveniles he protects them from the aggression of the females, because it has been shown that females will accept strange juveniles that have been chinned by the dominant male of their group whereas they will drive out others that have not been so marked (Mykytowycz, personal communication).

When social groups are being formed or re-established at the beginning of the breeding season those rabbits that are evicted from or fail to gain access to established territories may form new social groups, especially if the ancestral warren is not heavily populated, or if there are derelict warrens nearby that can be taken over and renovated. But when the population is dense there are usually a number of expatriate outcasts living solitary in the no-man's-land between warrens.

Myers and Poole studied the rabbits' ability to maintain their social groups though densely crowded. They put varying numbers of rabbits, in breeding condition, into three two-acre enclosures where there was a good pasture of clover and grass. The rabbits bred during two successive winters, but many of the young died during the summer. Table 3.04 shows that, no matter how intense the crowding, the rabbits did not vary the size of the social group; they responded to crowding by reducing the size of the territory.

Clearly the behaviour that causes rabbits to live in social groups of about eight is strongly ingrained in them. The selective advantage seems to come from being a member of the group *per se*; it is not confined to the dominants or the subordinates. Both are necessary for the persistence of the group; and natural selection has ensured the persistence of both sorts in the population partly by gearing dominance to tradition, age and weight and partly by not allowing either dominant or subordinate an exclusive advantage. Within the social groups Myers and Poole could find no evidence that dominant females contributed more or stronger progeny to posterity than the subordinates; but the dominant males seemed to have an advantage over the subordinates. Even so there seemed to be selection pressures to offset this advantage, notably the strong tendency of the females in a group to synchronize their reproductive cycles. This tended to spread paternity because it is difficult for the dominant male to attend more than one pregnant female at a time.

The territorial behaviour of the rabbit, unlike that of *Parus* or *Gymnorhina*, does not seem to result in the 'hoarding' of food against

Table 3.04. *The influence of the density of the population in an enclosure on the size and number of social groups (After Myers and Poole, 1959)*

Year	Enclosure	Number of adult rabbits	Number of social groups	Mean Number per group		Mean size of territory acres	Number of outcast males
				♂	♀		
1957	A	6	1	2·0	4·0	2·00	0
	B	10	2	2·0	3·0	0·88	0
	C	19	3	2·3	3·3	0·66	2
1958	A	38	6	2·5	3·8	0·35	2
	B	51	7	2·3	4·7	0·29	2
	C	49	8	2·3	3·3	0·33	0

emergencies for the benefit of the residents at the expense of the expatriates because the rabbits' territories are readily compressible. Indeed it was a common experience, before myxomatosis was introduced into Australia, to come across dense populations of rabbits starving because they had been overtaken by a drought. The most obvious advantages of belonging to a group are: (*i*) A group, by virtue of its ability to defend a cluster of burrows, can retain the same burrows throughout the breeding season or for several seasons, which allows the group to develop its burrows. The more elaborate the burrows are the better the protection that they afford against the weather and predators. A group can make a more elaborate system of burrows than a solitary rabbit. The more elaborate the burrows become the more attractive they are to strangers and the greater the need of a well-knit group to defend them. (*ii*) By living within their territory the rabbits become familiar with it, gaining confidence through familiarity, and hence greater security in seeking food and avoiding predators.

Perhaps these advantages to the member of a group are sufficient to explain the social behaviour of the rabbit in which case social behaviour in the rabbit is best seen as an adaptation that allows the rabbit to make better use of a resource (burrows) by improving its quality. In the ecological context there is not much difference between such a behavioural adaptation and a physiological adaptation such as the development of a caecum which allows the rabbit to make better use of fibrous food. The evolutionary principles that have been proved in the study of physiological adaptations are equally relevant to the evolution of behaviour (Ewer, 1968).

3.3. THE DISTRIBUTION AND ABUNDANCE OF RESOURCES

In section 3.11 I described how the intensity of an extrinsic relative shortage might depend on various outer strands of the ecological web: weather influencing the accessibility of food for *Thrips imaginis*; temperature influencing the accessibility of food for grouse; food for sheep influencing the accessibility of food for *Ixodes*; and so on. An extrinsic relative shortage of a resource may also be generated if the resource is consumed or destroyed by a different species as, for example, when a plague of locusts or rabbits eat the plants that might otherwise have served sheep.

This sort of interaction in the ecological web has often been discussed under the heading 'interspecific competition'. But 'competition' has been used so widely and so uncritically that it has become what Hardin (1956) called a 'panchreston'. By analogy with a 'panacea', which is a remedy that cures every ill, a panchreston is a word that explains everything. Hardin wrote: 'The history of science is littered with the carcasses of discarded panchrestons: the Galaenic *humours*, the Bergsonian *elan vital*, and the Drieschian *entelechy* are a few biological examples in point. A panchreston which "explains" all *explains* nothing.' I hope that we ecologists may soon be able to add 'competition' to this list of 'abandoned carcasses'.

Fortunately plagues of locusts or of rabbits are usually intermittent. But when this sort of interaction is persistent it may lead to the exclusion of a species from a place where it might otherwise thrive; this is especially likely if the resource is essential and has no substitute. Such an interaction has been nicely analysed by Reynoldson (1966) in an elegant investigation of the ecology of four species of triclads that live in freshwater lakes in Britain. All the species ate a variety of food; especially oligochaetes and arthropods; but *Polycelis nigra*, *P. tenuis* and *Dugesia lugubris* preferred oligochaetes whereas *Dendrocoelum lacteum* preferred crustacea; *D. lugubris* would also eat molluscs but *D. lacteum* would not. Reynoldson found that the abundance and diversity of food for triclads varied directly with the concentration of calcium in the water. Table 3.05 shows the relationship between the calcium content of the water, the quantity and diversity of the food and the abundance of the different species of triclads.

Table 3.05. *The distribution of triclads and their food in lakes of varying richness as measured by the concentration of calcium in the water. Total food was measured as biomass of standing crop ranging from 0·4 to 6·7 g collected per hr. The distribution of molluscs and crustacea were scored as the proportion of lakes in which Lymnea and Hydrobia or Asellus occurred: the range was from 20% to 83% for the molluscs and from 10% to 83% for Asellus. The abundance of triclads was scored as the number caught in an hour: the range was from 0 to 160 (After Reynoldson, 1966)*

Calcium, mg/l	<5 mg/l	5–10 mg/l	>10 mg/l
Food: molluscs	trace	+ +	+ + +
Food: crustacea	trace	+ +	+ + +
Food: total	+	+ +	+ + +
Polycelis nigra	75	40	35
P. tenuis	20	100	160
Dugesia lugubris	0	20	80
Dendrocoelum lacteum	0	20	60

In rich lakes, with more than 10 mg of calcium per l., where food is abundant and diverse the four species were often found living together and *P. tenuis* was usually the most abundant species. *D. lugubris* and *D. lacteum* were not found (with one minor exception) in lakes with less than 5 mg of calcium per l. where the food was obviously less diverse (Mollusca and Crustacea had virtually disappeared). Reynoldson thought that in these lakes the *Polycelis* species, being better adapted for living on the remaining sorts of food, caused a shortage of food for *D. lugubris* and *D. lacteum* which they could not tolerate, and could not alleviate by turning to an alternative sort of food. In 41 lakes with less than 5 mg of calcium per l., 27 contained *P. nigra* alone, 12 contained *P. tenuis* alone and in 2 both species were found in the same lake. Reynoldson could see no consistent difference in the lakes to account for the distribution of *Polycelis* spp. in them. But it is likely that these two species also differed in their requirements for food (though more subtly than they differed from *Dugesia* and *Dendrocoelum*) and that some lakes favoured one species and some the other; but in only two lakes was the food diverse enough to provide a 'food refuge' for the less fortunate one.

It seems likely that shortages of resources are more frequently caused by the destruction than by the consumption of a resource by a different species. Ritchie (1920) described how a population of red grouse *Lagopus scoticus* came to be excluded from a typical heather moor in Scotland where they had lived for a long time; and subsequently recolonized it. In 1892 a few gulls *Larus ridibundus* began to nest in the area. The owner of the moor protected them and they increased until by 1907 there were 3,000 birds nesting there. By this time the soil and vegetation and the concomitant animals had been changed enormously by the dung and the tramplings of the gulls. Pools of water formed where before there had been dry ground; the heather disappeared; rushes (*Juncus*) and docks (*Rumex*) dominated the new vegetation that replaced the heather and its associates. The resources of shelter and food that a heather moor usually offers red grouse dwindled and so did the grouse; eventually the shortage became intolerable and no grouse was left. Subsequently protection of the gulls ceased; their numbers began to decline; and by 1917 there were less than 60 pairs nesting in the area. At this time the heather was beginning to thicken up and spread again and the moor began to support a few grouse. In due course the area reverted to its original condition and looked like a typical heather moor once again, complete with a normal population of grouse.

Elton (1927) quoted a number of other examples including some in which the 'other species' in the interaction was man. Indeed man is the most destructive species of all. Examples of his destruction of the resources of other sorts of animals are too numerous to mention and too obvious to need mention. In North America the bison and the passenger pigeon are well known because they were spectacular (Hewitt, 1921). In Australia Calaby (1963) mentions six species of marsupials that he thought had recently become extinct. He thought that most, if not all of them, had died out because they had met with an intolerable reduction of resources as their habitat was destroyed by the clearing of land for farms or by the grazing of sheep. In many large areas the whole countryside has been changed: so greedy has this exploitation been.

However man's exploitation of the countryside has served to increase the supply of resources for some species; insects, rodents and other sorts of animals that are pests of crops, livestock and stored products are prominent in this category. I mentioned in section 1.1 that outbreaks of the grasshopper *Austroicetes* occasionally do a lot of damage to pastures in South Australia in the area shown in Fig. 1.01. Less than 150 years ago, before this area was settled by Europeans, it carried a woodland vegetation of *Eucalyptus* spp. in parts, and in other parts a scrub of *Acacia*, *Kochia*, *Atriplex* and other shrubs. The only places suitable for the breeding of *Austroicetes* would have been grassy glades which occupied only a small proportion of the total area. As the land was occupied the vegetation was cleared. In some areas the land was cultivated for wheat for a number of years and then converted to pasture; in other places pastures were developed from the beginning. But in either case the pastures contained much food for the grasshoppers and many places that were suitable for them to lay eggs in. Several thousand square miles were converted into breeding grounds for *Austroicetes* with a consequent enormous increase in their numbers (Andrewartha, 1943, 1944).

The musk-rat *Ondatra zibethicus* usually lives along the edges of shallow water where it builds burrows or 'lodges'. Marshes are good; so are drains if they are fed by tiles, especially if they run near fields that provide food. Streams also provide places for musk-rats to live, but according to Errington (1944, 1948) a stream becomes increasingly suitable the more it resembles a marsh. Fluctuations in the level of the water may cause fluctuations in the carrying capacity of the area.

Temporary changes in carrying capacity may be brought about by weather. The carrying capacity of large areas has been changed artificially by draining, damming or ditching in areas of marshland or other suitable territory. Dymond (1947) referred to the construction of dams, dikes and canals in the Saskatchewan River, Manitoba, which diverted water into dry marshes and thereby greatly increased the number of suitable places for musk-rats to live: the number of musk-rats in the area increased from 1,000 to 200,000 in four years.

With the four species of triclads that Reynoldson studied the first to disappear as the supply of food became more homogeneous were the two whose ecologies differed most from the two successful species. As the supply of food became even more homogeneous one or other of the two *Polycelis* spp. also disappeared. Reynoldson could not document the cause but there seems to be no reason to seek a different explanation. With the red grouse and other species for which the resource is destroyed rather than consumed, it is clearly the difference between the two species that makes one of them vulnerable. In section 7.2 I mention several examples of species that seem not to be able to co-exist for reasons not associated with resources; the explanation for their failure to live together lies in their essential differences rather than in their similarity.

In this perspective the 'competitive exclusion principle' which states that two species with the same ecology cannot co-exist (Hardin, 1960) seems trivial and the criticisms of it (e.g. Cole, 1960a, 1960b) seem to be just.

In section 3.12 I discussed the ways in which heavy exploitation of a resource by a dense population might change abundance into intrinsic relative shortage. In these circumstances the central strand in the ecological web which directly influences each individual animal's chance to survive and reproduce is the shortage of food; but the outer strand of the web which causes the shortage is the large number of other animals of the same species that are also seeking to use the resource.

Andrewartha and Browning (1961) proposed a diagrammatic method of representing the 'status' of an animal relative to its resources for all the conditions of relative shortage covered by sections 3.11 and 3.12 and for those conditions of absolute shortage covered by category B(*b*) of Table 3.01 (Fig. 3.03).

Figure 3.03 is like a map. The position on the map of any persistent population of animals is fixed: (*a*) by the proportion of the population that get enough of the resource to complete their life-cycles (i.e. by the severity of the shortage that they experience); (*b*) by the abundance of the animals relative to the resource (i.e. the proportion of the resource that is likely to be consumed). For example *Glossina* and *Ixodes* are placed near the bottom left-hand corner of the map because they both characteristically experience a severe extrinsic shortage of food; at Shinyanga *Glossina* eventually disappeared off the map at the bottom left-hand corner (section 3.11).

Species which experience a relative shortage are located towards the right when they characteristically consume a large proportion of the resource (intrinsic shortage) and to the left when they characteristically consume a small proportion of the resource (extrinsic shortage). A successful agent of biological control such as *Cactoblastis* or *Rodolia* (section 3.12) is likely to trace a characteristic pathway on the map. In Fig. 3.c3 this pathway is indicated by the three positions, and their dates, plotted for *Cactoblastis*.

Animals that remain abundant relative to a resource for reasons not related to the resource experience no shortage of that resource. Their place on the map is near the extreme top. *Eriosoma* is kept rare relative to its food by a predator, *Aphelinus* (section 3.12); it is surrounded at all times by an abundance of readily accessible food and it eats a very small proportion of it. *Cactoblastis* in 1926 was rare relative to its food because it had only recently been introduced to the area and had not had time to increase greatly. The sheep is kept numerous relative to its

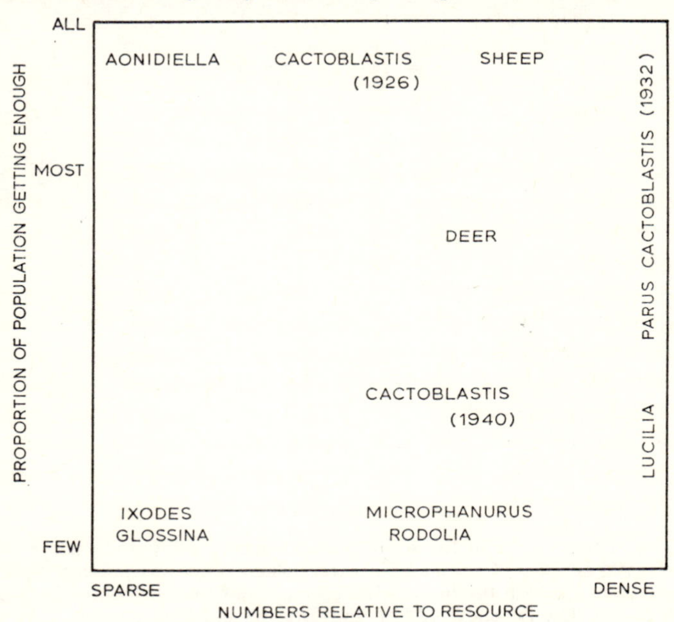

Fig. 3.03. Map of the 'status' of animals relative to their resources. For explanation see text. (After Andrewartha and Browning, 1961.)

food (but not so numerous that any go short of food) by the husbandry of the farmer. So it is placed in the top right-hand corner of the map. It is difficult to think of a natural population that would take this position.

Absolute shortages that develop in local situations as defined by categories A(*a*), A(*b*) and B(*a*) of Table 3.01 have no place on the map; such local populations might be imagined as sliding off the map in the bottom right-hand corner.

Interactions among the outer strands of the ecological web may influence the quality of a resource as well as its quantity or its distribution.

For example the Colorado potato beetle *Leptinotarsa decemlineata* in Montana burrows into the soil at the end of summer and remains there hibernating during winter. In light, sandy soil they may burrow to a depth of 14–24 inches, but in harder soils most of them are likely to be within the top 8 inches, concentrated just above the 'plough line'.

Mail and Salt (1933) measured the cold-hardiness of *Leptinotarsa* in the laboratory and concluded that all the beetles would die if they were exposed to $-12°C$ for even a few minutes; more than half would die

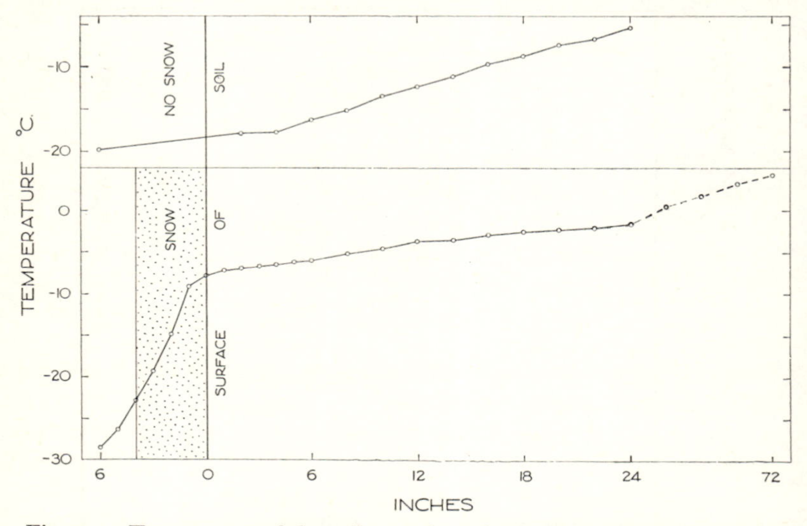

Fig. 3.04. Temperature of the soil at various depths below the surface: *upper curve*, in the absence of snow; *lower curve*, when snow was present. Note the steep gradient through the four inches of snow and the consequent less severe temperature below the snow. (After Andrewartha and Birch, 1954 – data from Mail, 1930.)

after several hours at $-7°C$; but most of them would survive for many days or weeks at $-4°C$.

They then measured the temperature of the air six inches above the ground, and in the soil down to a depth of six feet. The measurements were taken twice, once with and once without a cover of snow on the ground. The results are shown in Fig. 3.04. In the absence of snow a beetle would have needed to be at least two feet below the surface to have had much chance of surviving, but in the presence of snow most of the beetles that were 12 inches below the surface would probably have been safe.

The fox *Vulpes vulpes* is a predator of the rabbit *Oryctolagus cuniculus*. Because it is clumsy the fox rarely catches a rabbit on the surface, even a small one, unless the rabbit is decrepit from disease. The fox's style is to dig for nestlings. It can, with great skill, locate the nest from the surface; then it digs vertically downwards. Myers and Parker (1965) recorded the number of nests that had been dug out by foxes in different sorts of soil (Table 3.06).

Table 3.06. *Predation by foxes on rabbits making their burrows in different soils (After Myers and Parker, 1965)*

Soil	Number of warrens		Number of nests destroyed
	total	attacked	
Stony banks	829	0	0
Loam	871	66	78
Stony hills	806	11	12
Stony pediments	371	46	58
Desert loam	324	148	255
Sand dunes	3309	2339	2550

Table 3.06 shows that nests are more secure from the fox when the burrow is made in hard or stony soil.

4 *Components of environment: mates*

4.0 INTRODUCTION

Andrewartha and Birch (1954) recognized a component of environment 'other animals of the same kind' which they discussed under two major headings 'underpopulation' and 'crowding'. The disadvantages of living in a sparse population were discussed in relation to (1) the risk of not finding a mate and (2) the risk that some other component of environment might be less favourable for individuals in a sparse population. The disadvantages of living in a dense population were discussed in relation to (3) the risk of generating a shortage of resources, (4) the risk of stimulating the multiplication of predators by providing them with an abundance of food and (5) the risk that intraspecific aggression might adversely influence the physiological condition of animals living in a dense population thereby reducing their chance to survive and reproduce.

Only of items (1) and (5) can it be said that an animal's chance to survive and reproduce is *directly* influenced by the number of the other animals of its own kind that it is likely to meet. Items (2), (3) and (4) refer to ways in which the activity of some other component of environment might be modified. In such interactions other animals of the same kind appear as strands of the ecological web that are not part of the environment as I have defined it in chapter 2. These interactions are discussed, in their appropriate contexts in chapters 3, 5, 6, 7 and 8. Intraspecific aggression is discussed in chapter 5. The scope of this chapter is restricted to the risk of not meeting a mate in a sparse population and adaptations for reducing these risks.

4.1 SHORTAGE OF MATES

In sparse populations an animal's chance of reproducing increases as the number of potential mates increases, as illustrated in Fig. 4.01.

The sheep tick *Ixodes ricinus* has been established in northern England for centuries; but there are still some farms that are free from ticks although they seem to be quite well suited to them. It is known that, on occasion, the spread of the ticks from one farm to the next has been held up for decades by no greater barrier than a good sheep fence. It is also known that even after *Ixodes* has become established in a new area it has multiplied slowly, taking many years to reach the level of abundance that is characteristic of populations that have been established for, say, 50 years. Both these problems were studied by Milne (1950, 1951, 1952).

The ticks live for most of their lives in the vegetation near the ground but they do not feed or mate except on a sheep or some other animal.

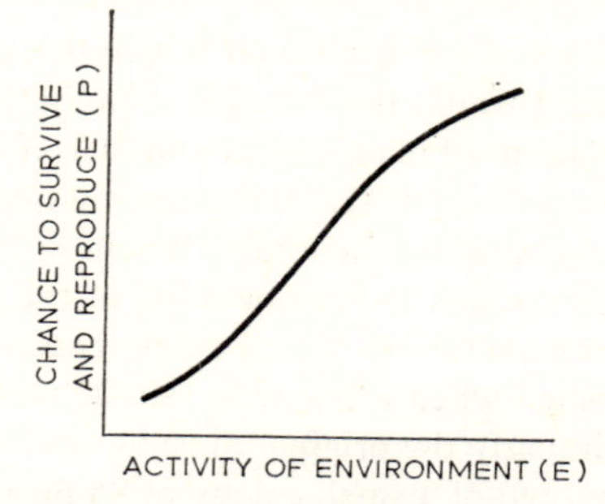

Fig. 4.01. An animal's chance of meeting a mate is enhanced when potential mates are more numerous. *E* stands for the activity of the environment (number of potential mates); *P* stands for the animal's chance to survive and reproduce.

At the appropriate time of the year the ticks emerge from the mat of vegetation and climb to the tip of a protruding grass-stem. If, during the week or so that the tick can stay alive in this position, a sheep brushes by, the tick will cling to the sheep. If a sheep picks up two ticks of opposite sexes they may come together and mate; they are not much good at crawling so their chance of meeting might be quite small if they happen to be picked up at opposite ends of the sheep's body.

Ignoring this last risk and making certain approximate assumptions, Andrewartha and Birch (1954) calculated the chance that a tick might have of meeting a mate. In a population in which the density was such that on the average each sheep picked up one tick, each one that was picked up by a sheep would have one chance in five of meeting a mate. (The chance is 0·2 and not 0 because by chance some sheep carry 2 or more ticks and some carry none, even though the average is one.) As the average number per sheep increased to 10 the average tick's chance of meeting a mate rose to 75%.

These figures do not take account of any other accidents such as the chance that the two sexes will be picked up at different times or on widely separated parts of the sheep. Also Milne estimated that on the average, a tick that is just beginning life has about one chance in ten of living to become mature. When all this is taken into account it is easy to appreciate that the occasional tick that is carried through a good sheep fence, by a straying badger or some other animal, would have virtually no chance of founding a colony. If the ticks are to have any chance of becoming established in a new area the original colony must be large; this happens when a whole flock of 'ticky' sheep is moved into a new area.

I do not know of any other species for which this principle has been so nicely worked out. I would think that it is likely to be commonplace among sparse populations of many sorts of animals if only we had a way to observe it. And it seems to me that the success of quarantine measures, which are aimed at preventing the establishment of pests in new areas, must often owe their success to the operation of this principle. On the other hand, the European rabbits which are now such pests in Australia first became established when a few wild rabbits were introduced into Victoria in 1859. Similarly the original introduction of the starling into North America consisted of a small colony of 80 birds which nevertheless established itself in the new area (Elton, 1958).

There have been occasions when ecologists have artificially caused a shortage of mates in order to reduce the numbers of an insect pest. The first occasion that I know of was when Glover *et al.* (1955) were able to exterminate *Glossina morsitans* from 280 square miles of bush in northern Rhodesia by clearing the scrub from small local features called Kasengas (section 1.2).

Knipling (1955) proposed a method which was subsequently used to exterminate the screw-worm fly *Caliroa hominovorax* from the small island of Curacao which has an area of 170 square miles. Large numbers of sterile males were persistently released into the natural population. The males had been reared in an insectary and sterilized as pupae by exposure to doses of X-rays which destroyed the sperm but did not otherwise impair the sexual vigour of the flies. The sterile males copulated with the wild females thereby stimulating them to lay eggs which were sterile. In Curacao weekly releases of 400 sterile males per square mile for six months (about four generations) was sufficient to exterminate *Caliroa hominovorax* from the island. Subsequently the same method was used on a larger scale, and with equal success, in Florida (Knipling, 1960). Since then this method has been demonstrated experimentally (but not yet used on practical scale) for the fruit-fly *Dacus tryoni* (Andrewartha *et al.*, 1967). This method seems certain to attract increasing attention from insect ecologists as the disastrous pollution of our own environment with the poisonous residues of insecticides continues.

A variation on Knipling's method was reported by Bateman *et al.* (1966). Males of the fruitfly *D. tryoni* are strongly attracted to any object that gives out the odour of cue lure, 4-(*p*-acetoxyphenyl)-2-butanone. Bateman impregnated pads of artificial board with cue lure and an appropriate poison, malathion, and distributed them densely through his study-area. Many males were killed, with the consequence that many females failed to copulate and the population was greatly reduced.

4.2 THE PREVALENCE OF SPARSENESS

Those species of animals that are obvious because they are abundant are unusual; probably they represent only a fraction of one per cent of all the species that are known. Darwin, in Chapter II of *The Origin of Species*, wrote: 'Rarity is the attribute of a vast number of species of all classes, in all countries. If we ask ourselves why this or that species is rare, we answer that something is unfavourable in its conditions of life; but what that something is we can hardly ever tell.' Elton (1938, p. 130) wrote: '. . . animal numbers seldom grow to the ultimate limit set by food supply and not often (except in some parts of the sea) to the limits of available space. This conclusion is also supported by the general experience of naturalists that mass starvation of herbivorous animals is a comparatively rare event in nature, although it does occasionally happen as with certain caterpillars that abound on oak trees in some years and may cause complete defoliation. With predatory animals it probably happens more often.' Smith (1935, p. 880) wrote: 'In general, attempts to determine abundance or scarcity have been based upon studies of species which are of economic importance only. The fact that the number of species which become sufficiently abundant to damage crops is relatively small, and that such species form only an insignificant fraction of the total number of phytophagous insects, is ignored.' A graphic picture of the rareness of certain species of scale insects relative to their food is given by Cockerell (1934), in the passage that I quoted in section 3.12. And, in general, the evidence would seem to suggest that rareness is the usual condition of most species of animals; even those species that are obviously abundant becoming rare in some parts of their distribution and perhaps in most parts of the distribution at some time.

Consequently the principles that govern sparse populations which have been discussed in this chapter (see also section 8.1) are widely relevant to the general theory of population ecology. And it is not surprising that diverse adaptations have arisen in response to the risks associated with membership of a sparse population.

4.3 ADAPTATIONS THAT INCREASE THE CHANCE OF FINDING A MATE WHEN NUMBERS ARE FEW

Cue lure (section 4.1) is an artificial substance that was first synthesized in the laboratory and subsequently found to attract males of the fruitfly *Dacus tryoni*. Whether cue lure has a counterpart in nature is not known. But lepidopterists have known for many years that the best way to catch males of certain species of moths is to expose a captive female on a warm evening with a gentle breeze. The males detect the pheromone that is given off by the female and follow it upwind from great distances. Pheromones that serve as sexual attractants are known for a diverse range of animals including many mammals (Wilson 1963).

Waloff (1946b) recorded the wanderings of many swarms of locusts, *Schistocerca gregaria*, in East Africa. The movements of the swarms were largely governed by temperature and wind. A characteristic response to rising temperature and wind made it likely that the locusts would be picked up by ascending turbulence and carried perhaps 2–3,000 ft above the ground where a strong wind (its likely direction depending on the season of the year) might blow them perhaps hundreds of miles before a descending turbulence would bring them to the ground again. There was no evidence that the wanderings led, except by chance, to favourable places for breeding; but when, in the course of their wanderings with the wind, the locusts arrived at a place that was moist enough, they would develop to sexual maturity, copulate and lay eggs. The advantages conferred on them by their gregarious behaviour, keeping them together until they were ready to copulate, is obvious. In other circumstances gregariousness may have other advantages, but it is a fact that many insects, fish, birds and mammals whose usual way of life includes long 'migrations' are gregarious.

In section 3.212 I described how the European rabbit formed small social groups and defended a territory, and how the males in the group defend the females against any foreign male. The American cottontail rabbit *Sylvilagus floridanus* forms similar social groups and the males 'defend' the females, driving away all other males. Although the group does not defend a territory as such they usually remain in a limited 'home-range'. This sort of social behaviour is developed further in many of the large ungulates. According to Ewer (1968):

> In many artiodactyls, seasonal movements of the herds in relation to changing food supplies makes the holding of a stable breeding territory impossible. This, however, need not abolish 'territorial' behaviour on the part of the males; instead of a definite area, the group of females itself is defended, regardless of the fact that they are constantly on the move. We have already had an example of this in the impala, where the area over which the animals feed is too large, in relation to the degree of cover, for its defence as a spatial unit to be possible. The same principle operates even more strikingly in the case of the blue wildebeast, *Connochaetes taurinus*. In the animals studied by Talbot and Talbot (1963) in Masailand, mating occurs when the animals are on the move from the open plains towards the bush. . . . The size of the breeding groups is highly variable and may be anything from two or three to over 150 individuals. If the group of females and yearlings is large, then two or even three males may share it and cooperate in defending it against others, although showing no animosity to each other.

It is Ewer's opinion that this behaviour ensures that 'all females breed while there is a selection of males'. Ewer also mentions other advantages pertaining to membership of a herd which may be independent of the density of the population; but a behaviour that ensures the breeding of every female is obviously a useful buffer against the risk of not meeting a mate when the population is sparse.

5 *Components of environment: predators and pathogens; aggressors*

5.0 INTRODUCTION

The conditions caused by pathogens (microorganisms and viruses) are called diseases; so are some of the conditions caused by metazoan parasites which are also smaller than their hosts; but an animal that is eaten by another animal that is larger (or not much smaller) than itself is said to be the prey of the larger animal which is called a *predator*. Predators usually kill and eat the whole prey; parasites and pathogens may eat without killing; they may merely eat or they may also produce toxins that weaken or kill the host. For the purpose of ecological analysis the activities of all three groups are best placed in one category which needs a name. 'Predation' seems to be the most general one that emerges from common speech about predators, parasites and pathogens; so I shall call them all predators unless I want to make a narrower distinction.

In contrast to the broad usage for 'predator' a very narrow meaning has been reserved for *aggressor*. I do not, in this chapter discuss the aggression that goes with the defence of a territory. I regard territorial behaviour as an adaptation that has evolved in animals in relation to their utilization of resources; so I discuss it in chapter 3. Similarly the aggressive social behaviour found, for example, in a male cottontail 'defending' a group of females is an adaptation properly discussed in chapter 4. In this chapter I am interested only in the special sort of aggression that develops in crowded populations especially of small mammals and is said either to alter the animal's physiological condition so it becomes less viable or less fecund, or else it is said to act directly on the animal, reducing its chance to survive and reproduce, without necessarily altering its physiological condition.

Because the activities of predators, parasites, pathogens and aggressors are broadly parallel, and can be represented by a generalized curve like the one in Fig. 5.01 I have recognized only one component of environment to embrace them all.

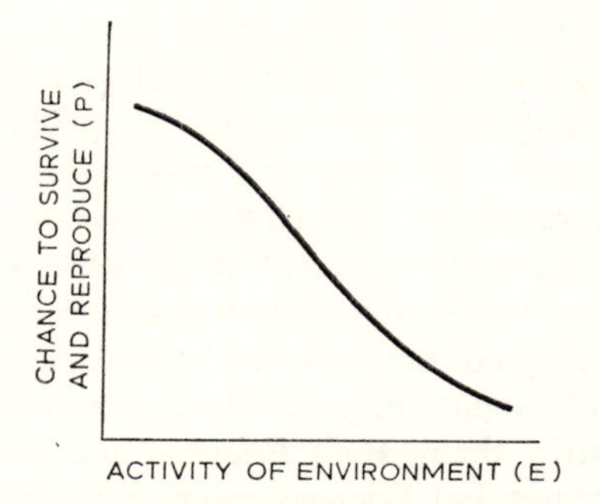

Fig. 5.01. An animal's chance of survival is reduced when predators, pathogens or aggressors are numerous. *E* stands for the activity of the environment (predators, etc.); *P* stands for the animal's chance to survive and reproduce.

5.1 PREDATORS

The activity of a predator (using the term in its broad sense) depends on its supply of food and all the other components of its environment. If the predator is restricted to one species of prey (as many insects and other invertebrates are) it is called an *obligate* predator. If it readily eats more than one species of prey it is called a *facultative* predator.

5.11 *The 'biological control' of insect pests*

About 170 years ago the scale insect *Icerya purchasi* was accidentally introduced into southern California from Australia. By 1880 it had spread and multiplied so much that it threatened to destroy the citrus industry. In 1888 a colony of 129 ladybirds *Rodolia cardinalis,* from Australia, were established inside a cage that had been built over an orange tree; the tree was carrying a dense population of *Icerya.* By April 1889 the ladybirds had increased greatly and had destroyed nearly all the scale insects. The cage was removed, and within three months *Rodolia* had spread through the orchard destroying nearly all the *Icerya* that were in it. This success was repeated elsewhere in southern California and ever since there have been few *Icerya* in this area. Smith (1939) explained this result in terms of the high powers of dispersal of the predator relative to its prey and also the high rate at which *Rodolia* multiplied. They multiplied rapidly not only because their food was plentiful and readily accessible but they were also favoured by other components of environment especially weather.

The woolly apple aphid *Eriosoma lanigerum* was accidentally introduced into Western Australia about the middle of last century. Despite regular and frequent spraying with insecticides aphids remained abundant until the encyrtid *Aphelinus mali* was introduced from North America about 1925. Shortly after *Aphelinus* was established *Eriosoma* became quite scarce and has remained so ever since. From midsummer through winter one may search diligently in the orchards with little chance of finding an aphid. But a few survive and usually, during spring, small colonies can be found, here and there, multiplying briefly before the predators destroy them. Both *Eriosoma* and its predator remain dormant during winter, but *Eriosoma* resumes its activities at a slightly lower temperature than *Aphelinus*. Hence it is able to breed for a brief period during early spring.

The ladybird beetle *Rhizobius ventralis* was introduced into California about 1895 in the hope that it might 'control' the scale insect *Saissetia oleae*. It became established and has persisted ever since, but from the economic point of view it has been substantially a failure. This is because the prey have considerably greater capacity for dispersal than the predators. None of the prey is concealed, in any physical sense, from the predators. Those individuals that escape being eaten, and live long enough to grow up and contribute progeny to the next generation do so, not because they are less vulnerable than their fellows, but purely because they are more fortunate: no predator happened to come in time to the place where they were living. In considering such examples, the emphasis is properly placed on the relative powers of dispersal of predator and prey (section 10.44).

The limit to the abundance of *Rodolia*, *Aphelinus* and *Rhizobius* is set by the relative shortage of food which is of their own making, i.e. it is an intrinsic shortage (section 3.12). With *Rodolia* and *Aphelinus* the shortage is severe, and both the prey and the predator persist as extremely sparse populations because the predators can out-breed and out-disperse the prey.

They can outbreed their prey because no component in their environments, other than the relative shortage of food, limits them. They are successful agents of biological control. From the perspective of the ecology of the prey one can say that *Icerya* or *Eriosoma* is kept sparse by the heavy pressure maintained by its predator; and one need not look

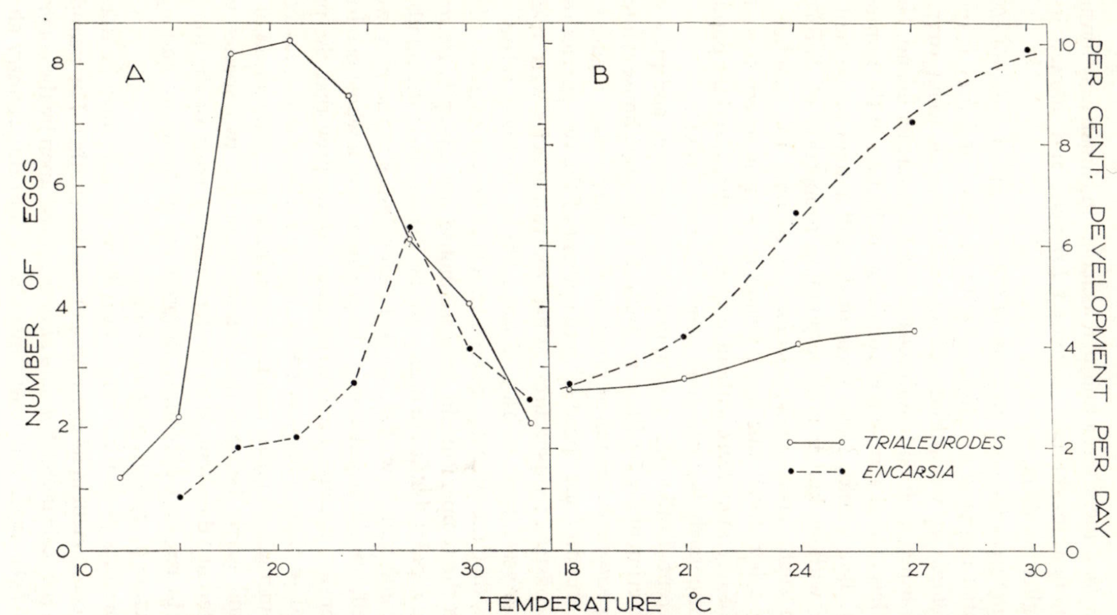

Fig. 5.02. The influence of temperature on A, the daily rate of egg-production, and B, the speed of development of a predator *Encarsia formosa*, and its prey *Trialeurodes vaporariorum*. Note that the relative advantage possessed by the prey at low temperature disappears at high temperature. (After Andrewartha and Birch, 1954 – data from Burnett, 1949.)

to the outer strands of the ecological web except to say that none is limiting.

With *Aphelinus* there was a hint that a change in climate, say towards a cooler summer, might have favoured *Eriosoma* and allowed it to maintain higher numbers.

With *Metaphycus helvolus*, predator of the scale insect *Saissetia oleae*, the differential response to temperature of predator and prey is more striking. *Metaphycus* was introduced into California in about 1937. It readily established itself both in the coastal area and farther inland. Near the coast it has caused a large and consistent reduction in the numbers of *Saissetia*. But inland, frosts occasionally kill a high proportion of *Metaphycus* without killing a corresponding proportion of *Saissetia*. In between such catastrophes, *Metaphycus* reduces the numbers of *Saissetia* to a low level; but for a period after each severe frost the numbers of *Saissetia* increase without any serious check, and the population may temporarily become quite dense (Clausen, 1951).

The differential influence of temperature on predator and prey was analysed experimentally by Burnett (1949) for the greenhouse whitefly *Trialeurodes vaporariorum* and its predator *Encarsia formosa*. Figure 5.02 shows the fecundity and speed of development of the two species over a range of constant temperatures. When the two species were reared together in a greenhouse at 18°C *Encarsia* bred so slowly relative to *Trialeurodes* that the predator offered no check to the increase of its prey. When the temperature in the greenhouse was 24–27°C *Encarsia* increased so rapidly that it ate nearly all the *Trialeurodes*; the numbers of both became quite few (Fig. 5.03). Local colonies of the prey were exterminated, but the population in the greenhouse as a whole persisted at a low density, because there were always some prey that the predators failed to find. This was partly because *Trialeurodes* was constantly colonizing new areas, and thus remaining one jump ahead of its predator, and partly because the behaviour of the two species was slightly different: they tended to prefer different parts of the tomato plants and to respond differently to gradients in humidity and light (see also section 9.12, Figs. 9.03 and 9.08).

The failure of a predator to press heavily on its prey may be due to some other component in its environment besides weather. For example, in Australia the scale insect *Saissetia oleae* is preyed upon by the encyrtid *Metaphycus lounsburyi*, which was introduced into this country about 1925. It seems to have had little influence on the abundance of *Saissetia*, and this has been attributed to the presence of *Quaylea whittieri*, which

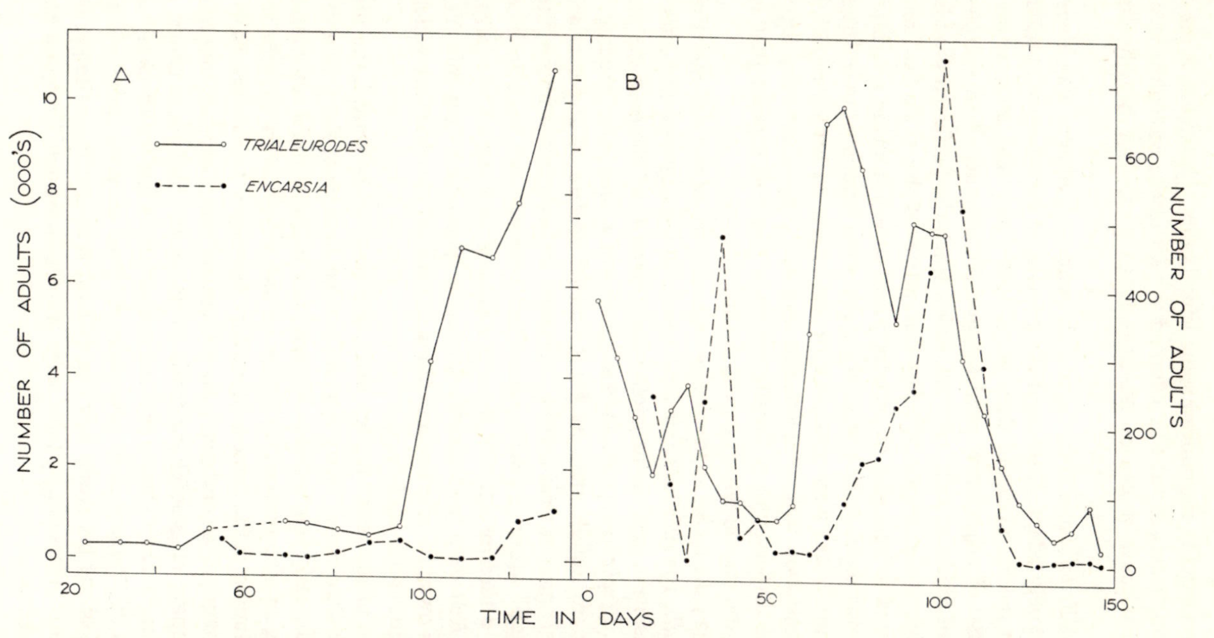

Fig. 5.03. The trends in the numbers of the predator *Encarsia formosa* and its prey *Trialeurodes vaporariorum* when they were reared together in a greenhouse where there were four tomato plants spaced several feet apart. In A the temperature was 18°C; in B, 24°C. (After Burnett, 1949.)

had been accidentally introduced some time before. *Quaylea* is a predator of *Metaphycus*; it must be counted as part of the ecological web of *Saissetia* because it causes *Metaphycus* to press less heavily on *Saissetia*.

Aphelinus, *Rodolia* and *Encarsia* are all obligate predators with their life-cycles closely adapted, through the prolonged operation of natural selection, to the life-cycles of their prey. Such close adaptation is not to be expected of a facultative predator, and the presence of an alternative source of food to tide it over the period when the preferred food is not available may be important. The caterpillars of the moth *Levuana irridescens* used to be a pest of coconut palms in Fiji. During a serious outbreak it was usual to find virtually every palm defoliated over large areas. The outbreaks were extensive because *Levuana* breeds rapidly; and the devastation was complete because the moths show very little tendency to disperse; they usually lay their eggs on a palm that is adjacent to the one on which they grew up.

About 1925 the tachinid *Ptychomyia remota* was introduced into Fiji (Tothill *et al.*, 1930). The adult *Ptychomyia* lays its eggs on the caterpillars of *Levuana* and the maggots which hatch from the eggs eat the caterpillars. *Ptychomyia* possesses the qualities of a 'successful' predator: it breeds more rapidly than its prey; and its powers of dispersal are incomparably greater than those of *Levuana*. Its success was immediate and thorough: within a year *Levuana* had become, and has remained ever since, a rare insect. Not even the famous example of the introduction of *Rodolia* into California was more spectacular than this. Yet *Ptychomyia* owed its effectiveness to one additional circumstance which was not required for *Rodolia*.

Only caterpillars that have reached a late stage in the third instar are suitable for *Ptychomyia* to lay eggs on. In the absence of the predators, or during the early stages of an outbreak, it is usual to find a substantial overlapping of generations; all stages of the life-cycle are present together, including the advanced caterpillars that serve *Ptychomyia*. In the declining stages of an outbreak, when there are not many *Levuana* left, the overlapping of generations tends to disappear, and there comes a time when there are no larvae present in a suitable stage for *Ptychomyia*. The adults of *Ptychomyia* live for no more than 10 days; quite a large proportion of the predators may die without reproducing if the period when no caterpillars of a suitable stage are present should be prolonged for several weeks.

On one island where no alternative food existed for *Ptychomyia* this sequence of events caused the numbers of *Ptychomyia* to fall so low that *Levuana* was able to increase greatly again before the predators 'caught up' with them. Elsewhere in Fiji two other species of caterpillars, *Plutella maculipennis* and *Eublemma* sp., that live in the 'bush' adjacent to coconut plantations serve as alternative prey for *Ptychomyia*. The predators maintain their numbers on these alternative prey during emergencies, returning to *Levuana*, which they prefer, when *Levuana* become available again. These species which serve as alternative prey for *Ptychomyia* are important animals in the ecological web of *Levuana*; in their absence, but with *Ptychomyia* present, *Levuana* might be less abundant in Fiji than it was before 1925, but certainly it would be much more abundant than it is now.

I think that many predators that press heavily on their prey follow the general pattern of *Ptychomyia*; that is, they are facultative predators with a strong preference for a particular prey but they can maintain themselves, during times of scarcity, on alternative species that also abound in the area. The converse was true of the facultative predators (birds) of the grasshopper *A. cruciata* (section 8.1; see also section 5.12).

The failures in biological control have greatly outnumbered the successes, but the successes have sometimes been spectacular, leaving no room for doubt that the newly-introduced predator has exerted a dominant influence on the abundance of the prey. The predator that is called 'successful' in this context may be thought of as one that outbreeds and outdisperses its prey. Conversely the 'failure' is one that breeds more slowly than its prey, either because it is not good at finding food even when there is plenty, or because some other component of environment retards its rate of increase relative to that of its prey. With *Encarsia* low temperature (18°C) would do this; with *Metaphycus lounsburyi* it was a predator. Failure to find more than a small proportion of the prey even when they are abundant may be due to a variety of causes; there was a high proportion of *Saissetia* that were never found by *Rhizobius* because the prey was very much better at dispersing than the predator.

5.12 *Predators of vertebrates*

The chief predator of the musk-rat *Ondatra zibethicus* in North America is the mink *Mustela vison*. Each year, as the breeding season approaches, a large proportion of the population is driven out from the familiar country where they have grown up, or at least have been living during the winter. The number of these outcasts is determined by the behaviour of the musk-rat, apparently independently of the presence of predators in the area. The musk-rats that are driven out are likely to be eaten by minks and other predators. But those that remain behind, established in a good 'lodge' and living in a familiar 'home-range', are very much more secure. Errington (1943) considered that this territorial behaviour of the musk-rats largely determined the size of the population; and the predators merely ate those that had already been determined as 'surplus' by their own fellows. And Errington (1946) considered that this principle applied generally to species that manifest 'territorial' behaviour (section 3.212).

Some of the larger ungulates seem to be able to protect themselves from even the canids which Errington classed as the most 'sagacious' and destructive predators except man. They can protect themselves because they are large and because they are organized as a herd. Consequently predators, except man, may have little influence on their chance to survive and multiply, while their numbers remain above a critical threshold (section 8.1).

Some of the smaller ungulates, antelopes, goats, etc., seem to be more susceptible to predators. Some of them live in inaccessible places, as the goats live in craggy hills. Others are more secure when they can form very large herds.

Errington (1946) in his review of predators among vertebrates concluded: (*a*) Predators are most likely to be influential in determining the abundance of the prey when the prey is a reptile or a fish, or among higher groups when the predator is one of the highly skilful canids (dogs, fox). (*b*) Predators are least likely to be influential when the prey manifest 'territorial' behaviour.

The European rabbit is territorial (section 3.212). In Australia the rabbit is the prey of two exotic mammals, the fox and the feral domestic cat, and a number of species of indigenous raptorial birds. Despite the territorial behaviour of the prey the predators sometimes live quite well off immature rabbits. None is much good at taking an adult. The breeding season lasts for two to six months during winter; so for the rest of the year the predators must seek alternative foods which are usually scarce or hard to come by. So the predators rarely become numerous enough to press heavily on the rabbits – and then only if, through some other cause, the rabbits have become few (compare the situation described for *A. cruciata* in section 8.1). With the musk-rat and other territorial species the casting out of the surplus is also highly seasonal.

5.2 PATHOGENS

5.21 *The biological control of rabbits by myxomatosis*

The myxoma virus which causes benign tumours in the American rabbits *Sylvilagus* spp. and which has always been endemic in them, has been known to be lethal to the European rabbit since 1896 (Sanarelli, 1898). For a few years after the disease first became established in Australia (about 1952) it consistently killed 99% or more of the rabbits that were infected. Fenner and Ratcliffe (1965) documented the history of myxomatosis in Australia. In the first attempt to introduce myxomatosis into a natural population of rabbits in Australia, Bull and Mules (1944) succeeded in inoculating many rabbits which usually transmitted the disease to other members of their own social group or of other social groups in the same warren (section 3.212). But the disease usually failed to spread beyond the warren into which it had been artificially introduced. Myxomatosis is known to be spread chiefly by bloodsucking arthropods. In the area where Bull and Mules worked, stickfast fleas *Echidnophaga myrmecobii* were numerous but mosquitoes were scarce. When they eventually abandoned the attempt Bull and Mules attributed their failure to the absence of winged vectors and suggested that myxomatosis might spread widely and become an effective agent of biological control in a place where there were many rabbit-biting mosquitoes.

The next attempt was made on a grand scale by a team of ecologists directed by F. N. Ratcliffe. Five areas were chosen close to the River Murray where the rabbits were numerous and where, in normal weather, it would be reasonable to expect many mosquitoes. Many rabbits were inoculated, but the results seemed to be little different from those reported by Bull and Mules. Fenner and Ratcliffe (1965) summarized the result of the season's work:

> By the beginning of December, myxomatosis had apparently died out everywhere except at Rutherglen, where the infection was fading towards extinction. The Murray Valley trials had been thorough: they had been carried out at a number of sites where the rabbit population was dense, and sometimes very dense indeed. Aedine mosquitoes had been reasonably abundant at Coreen; and at Balldale, where they were common, one of the three species collected had previously been shown by Bull and Mules (1944) to be capable of transmitting myxomatosis under laboratory conditions. At that juncture therefore the 1950 trials provided what seemed a clear confirmation of the conclusion reached by Bull and Mules

seven years earlier, i.e. that myxoma virus held little promise of being of practical value in rabbit control.

But even as this conclusion was being reached there was beginning a massive and spectacular epizootic (probably sparked off by a few infectious mosquitoes blown from the study-areas to regions further north where vectors were abundant), of which Fenner and Ratcliffe wrote: 'for scale and speed of spread [it] must be almost without parallel in the history of infections'.

The extent of the epizootic is shown in Fig. 5.04. During this first great epizootic wherever myxomatosis extended it infected most of the rabbits in that place; and it killed more than 99% of those that contracted the disease. Fenner and Ratcliffe commented on the map in Fig. 5.04:

Two things stand out. The first is the sheer size of the event recorded. The area over which the virus was dispersed – albeit patchily in some regions – measured nearly 1,000 miles from south to north and 1,100 miles from east to west. Apart [from a few outlying areas where inoculation had been artificial] the map is a record of the natural spread of the infection.

The second thing that stands out is the dominating importance of the Murray–Darling river system in the distribution of the myxomatosis activity. As we realized when we had learned something of the insect vectors, this was due to the fact that epizootic transmission had depended on summer mosquitoes that breed in permanent water, notably *Culex annulirostris*. The unusual rainfall distribution during 1950 was reflected in the great difference in the extent of disease activity in the southern and northern portions of the river system. In Victoria and southern New South Wales, 1950 was a year of sub-average rainfall, and by midsummer the countryside in general was very dry. In marked contrast, the northern parts of New South Wales and most of Queensland received record-breaking precipitations, resulting in flooding on an almost unprecedented scale in the northern part of the epizootic area. The last flood rains had occurred in November; but their effects in the form of abundant surface water lingered for at least a couple of months.

The spectacular epizootic depicted in Fig. 5.04 was caused by the extraordinary abundance of the vectors *Culex annulirostris* and the favourable turbulence and winds that dispersed them so widely. As Fenner and Ratcliffe pointed out in the passage quoted above this was an accident of the weather at that time. With more normal weather *Anopheles annulipes* might have been abundant earlier in the season and Ratcliffe's attempts to establish myxomatosis might not have failed – but then we might have had to wait a long time for an unequivocal

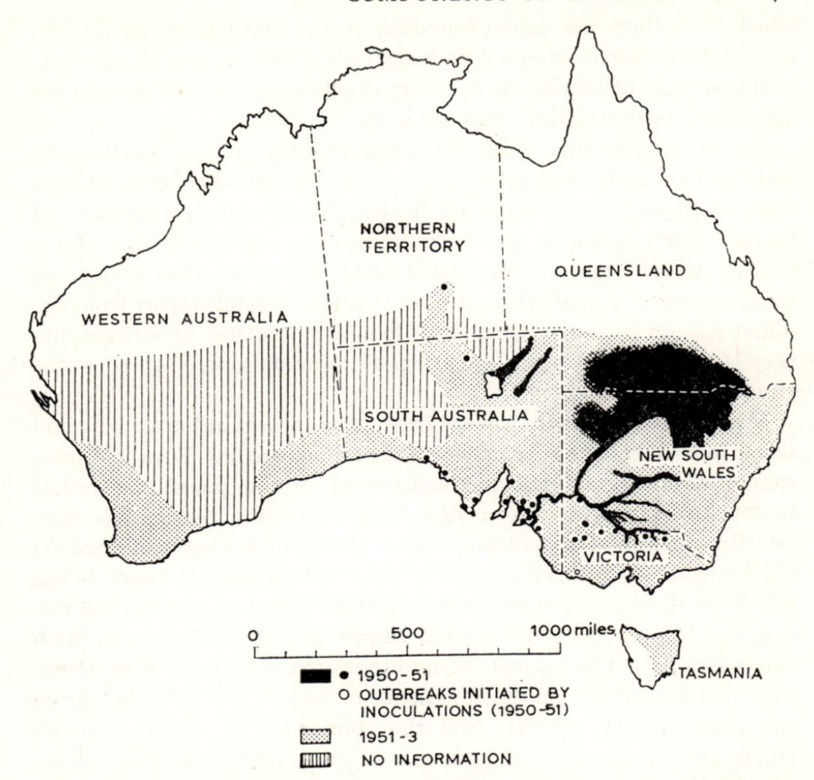

Fig. 5.04. Spread of myxomatosis in Australia during the three years following its escape, in 1950, from a test site near Albury. The isolated occurrences in 1950–51 ranged from single cases to outbreaks covering a few square miles. The two in central Australia, north and west of Lake Eyre, are included in the first season's dispersal although they were not reported until August and September 1951. (After Fenner and Ratcliffe, 1965.)

demonstration of the overwhelming importance of vectors; and for such clear evidence that no practicable increase in the density of the rabbit population can make up for a scarcity of vectors.

Although a variety of species of mosquitoes, gnats, fleas and other bloodsucking arthropods may transmit myxomatosis the two vectors of outstanding importance in eastern Australia are *Culex annulirostris* and *Anopheles annulipes*. They owe their importance to (*a*) their ability to breed in abundance in the sort of temporary waters that are likely to be abundant after rain; (*b*) their behaviour in feeding at dusk and during the night; (*c*) their behaviour in seeking to shelter in cavernous places which leads them into rabbit burrows; (*d*) their preference for feeding as adults on rabbits, coupled with their ability to support themselves, in the absence of rabbits, on a variety of other sources of food that are abundant, notably cattle, sheep and men.

For this reason the adults of these mosquitoes are not likely to go short of food even when rabbits are scarce. On the other hand, there is a persistent shortage of waters for the larval stages which is ameliorated by rain falling seasonally each year; but because the incidence of rain is variable the shortage is ameliorated more some years than others. So the distribution and abundance of the vectors depends (apart from the initial pattern of watercourses laid down by accidents of topography) very largely on the seasonal incidence of rain, and only slightly on the abundance of rabbits.

Myxomatosis normally persists between outbreaks at low density and seems to be able to do so even where the rabbits are few. But the disease cannot multiply and become widespread in the population unless vectors are numerous, no matter how numerous the rabbits may be. Presumably, extrapolating from the epidemiology of malaria (McDonald, 1957) there may be a critical low density of hosts, below which the disease is slow to spread even with plenty of vectors; but this question has not been investigated empirically for myxomatosis. With this one proviso it seems that the multiplication and spread of myxomatosis depends largely on the abundance of vectors and only slightly on the abundance of rabbits. And the abundance of vectors depends chiefly on the amount and incidence of rain, making breeding places for larvae.

Rainfall is variable, but over most of the distribution of the rabbit there is enough rain to maintain the numbers of vectors so high that myxomatosis continues to press heavily on the rabbit. The myxoma virus is a valuable agent of biological control. But it is no longer so spectacularly effective as at first, because the virus is evolving towards lower virulence, and the rabbit is evolving towards greater resistance. Will the end-point be similar to the relationship that already exists in North America between the myxoma virus and rabbits of the genus *Sylvilagus*? Fenner and Ratcliffe predict a different end-point; but this is an hypothesis the testing of which cannot be unduly hurried.

5.22 *The biological control of insect pests*

Entomologists have long been seeking pathogens to use as agents of biological control of insect pests and this subject now has an extensive literature going back more than 70 years. There have been few successes and none so successful or spectacular as myxomatosis. Attention was first focused on fungi because of several reports that outbreaks of locusts had come to a catastrophic end through fungal disease. At one time high hopes were held for the fungus *Beauvaria bassiona* against the bug *Blissus leucopterus*. But after Billings and Glenn (1911) had fully documented the shortcomings of *Beauvaria* as an agent of biological control, and it was realized that most entomophagous fungi were likely to share the same disadvantages, interest in fungi as agents of biological control has waned. Briefly the disabilities that Billings and Glenn attributed to *Beauvaria* were that it could multiply only during spells of weather so humid (and so continuously humid) that they were unlikely to occur often; and between outbreaks the fungus was scarce but ubiquitous so that there was nothing that an entomologist might do to increase its distribution or its abundance. Most entomophagous bacteria seem to be equally exacting and therefore unpromising as agents of biological control. An exception is *Bacillus popilliae* which causes the milky disease in the soil-inhabiting larvae of the Japanese beetle *Popillia japonica*. By virtue of highly-resistant spores which remain infective in the soil for a long time this pathogen manages to press fairly heavily on its host even when the population of the host is not dense (Beard, 1945).

Many entomophagous viruses are known and some seem to have an important influence on the numbers of their hosts.

According to Bird (1953, 1955) the sawfly *Neodiprion sertifer* is a serious pest of pine in Europe and Asia. During an outbreak extensive areas of pine trees may be defoliated. An outbreak usually ends abruptly with a catastrophic death-rate among larvae. A polyhedral virus, specific to *N. sertifer*, has been recognized as one of several causes for the collapse of an outbreak. *N. sertifer* was accidentally introduced into North America before 1925; by 1948 it had spread widely and become a serious pest; up to this time no one had found *N. sertifer* in North America infected with polyhedral virus. In 1949 Bird imported from Sweden several larvae that had died from polyhedrosis and established the virus in natural populations of *N. sertifer* in Canada. It is now a common practice to spray the pine forests with aqueous suspensions of polyhedra because this method has proved efficient and economical for controlling outbreaks of the sawfly.

The virus persists naturally in the forests and in the absence of spraying severe epizootics develop on dense populations of *Neodiprion*.

But the virus is sensitive to the density of the population of its host and at moderate densities (Bird estimated the threshold at about five egg-masses per tree) the sawfly is likely to defoliate a young tree before the virus catches up. The dispersal of this particular virus has not been fully documented but the severe disadvantage experienced by the virus when its host is not numerous may be related to the dependence of the virus on rather unlikely accidents for its dispersal. This is true of *Borrelina campeolis*, a polyhedral virus specific to *Colias philodece eurytheme*, which was studied by Thompson and Steinhaus (1950).

A polyhedrosis in *Colias* was first reported by Wildermuth (1911). Since then the disease has been much studied because *Colias* is a serious pest of alfalfa in California. There is no doubt about the capacity of *Borrelina* to bring an outbreak of *Colias* to a catastrophic end. Michelbacher and Smith (1943) recorded that during an interval of three days in midsummer (June) the number of living caterpillars declined from 14,000 per 100 sweeps of a net to 40 per 100 sweeps. On the first occasion they noticed a few diseased caterpillars; three days later dead caterpillars could be seen in their thousands. But it is doubtful whether *Borrelina* often prevents an outbreak from developing.

Thompson and Steinhaus (1950) found the distribution of the caterpillars with respect to alfalfa and the virus with respect to caterpillars patchy in the extreme. It was unusual to find a high proportion of caterpillars infected except in a dense population, but occasionally dense populations were free or nearly free of the disease and sparse populations heavily infected.

They showed the disease to be highly infectious. Experimentally they were able to infect larvae in various ways. (*a*) They ground up a dead caterpillar with some soil and allowed the infected soil to drift in a current of air past a clean plant. (*b*) They dipped leaves in water taken from an irrigation channel. (*c*) They allowed ants that had been feeding on a dead caterpillar to crawl over a clean plant. (*d*) They allowed a small parasitic wasp to oviposit in a diseased caterpillar and then crawl over a healthy one. (*e*) They placed the cut stem of a shoot of alfalfa in water in which a diseased caterpillar had been macerated. All these experiments gave high rates of infection.

It is easy to imagine how, even from a sparse inoculum, the disease might spread rapidly through a dense population of caterpillars. Probably this is its usual condition, which would explain why sparse populations rarely show a high incidence of the disease; but by chance, inoculum might be absent or virtually absent from an area where the population of *Colias* was dense. On the other hand a dense inoculum, brought in perhaps by heavily-contaminated irrigation water, might cause a high rate of infection in a sparse population; but this, according to the observations of Thompson and Steinhaus, must rarely happen.

5.23 *The activity of pathogens*

In general, pathogens seem not to press heavily on their hosts as some predators press heavily on their prey. If a pathogen is a fungus or a bacterium it may be so sensitive to unfavourable weather that it rarely has the opportunity to multiply strongly relative to its host. If the pathogen is a virus it may be so sensitive to the density of its host (i.e. it is very sensitive to a relative shortage of food) that it can thrive only in very dense populations of its host. Pathogens that have been living for a long time with a vertebrate host tend, by natural selection, to become benign as myxoma with *Sylvilagus* or measles with man. Myxomatosis in the European rabbit was exceptional because it had been introduced to a new host in a new continent. Also it is transmitted by a vector. Myxoma is not sensitive to the density of the population of its host (at least not in ordinarily sparse populations); neither is it sensitive to weather in its own environment. It is sensitive to weather in the ecological web of its vectors. But in Australia, by chance, over most of the distribution of the rabbit, weather is favourable enough to maintain dense populations of vectors at most times.

5.3 AGGRESSORS

The idea of aggressors as a component of environment arose in the first place by the convergence of two lines of work largely associated with the names of D. Chitty and J. J. Christian. In 1936 D. and H. Chitty began to study the ecology of several natural populations of voles *Microtus agrestis* living near Lake Vyrnwy in Wales. The work was interrupted in 1939 and resumed during 1945–59. The voles in any one local population seemed to reach a 'peak' density (of say several hundred per acre) about once every four years and then to 'crash'. They usually declined during one summer (which is the normal time for breeding) down to such low numbers that it would be difficult to say with any assurance that there were more than five or six per acre. From this low density the population would gradually increase during two or three years to another peak which would be followed immediately, or perhaps after a year of sustained high numbers, by another crash.

Characteristically the adults that were present during the spring and summer when numbers were at their peak would be on the average heavier than usual; and the immediate cause of the crash seemed to be poor viability of the juveniles that had overwintered and low fecundity of the young adults.

It seemed clear that weather was not directly causing the sudden decline in numbers because occasionally (though not typically) local populations adjacent to each other were found fluctuating out of phase. After careful research, predators, parasites and pathogens were also excluded. Voles and the other microtine rodents that have been chiefly studied in this connection are essentially herbivores. Although there are unusual records of the animals denuding an area of food, especially locally, it is usual for the crash to occur while there still remains, to a man's eye at least, abundant food. So, in formulating his hypothesis Chitty excluded food, along with weather, predators, parasites and disease from the likely causes of the decline.

Chitty (1967) argued that because all the other components of environment had been eliminated there remained only what Andrewartha and Birch (1954) had called 'other animals of the same kind'. In a dense population each vole has its chance to survive and multiply reduced by the aggression of its fellows. Moreover voles (not being born equal in this regard) may vary in their aggressiveness and their response to aggression. So in a dense and growing population the aggressive ones may leave relatively more progeny behind than the mild ones. Consequently, after a time, the population might contain mostly individuals that had been selected for their ability to withstand aggression. But the fact that the population has been increasing freely for a number of generations suggests that selective pressure by other sorts of risks may have been relaxed. So Chitty (1967) postulated, 'According to hypothesis the animals present in stationary or declining populations have been selected for their superior ability to withstand mutual interference. Animals in expanding populations are assumed by contrast not to have been so selected, but to be better fitted to withstand other hazards of their environment.' It is not clear what might be important among the 'other hazards of their environment' but Chitty discussed the hypothetical possibility of weather, acting directly on the animals (which he considered unlikely), or indirectly on the 'habitat' so altering the intensity of crowding and hence the rate of increase or decrease in the population.

According to Chitty's hypothesis this sequence of events might generate a feed-back and so keep the population between certain bounds; it is envisaged as a 'self-regulatory mechanism'. The feed-back is illustrated in Fig. 5.05 which I have copied from Chitty (1967). So far as I know this is the most modern version of Chitty's hypothesis which has been refined, as additional empirical evidence has accumulated, several times since it was first put forward by Chitty (1952).

In its original form, Chitty's hypothesis had been influenced by the work of Green and Evans (1940) and Green and Larsen (1938) on the

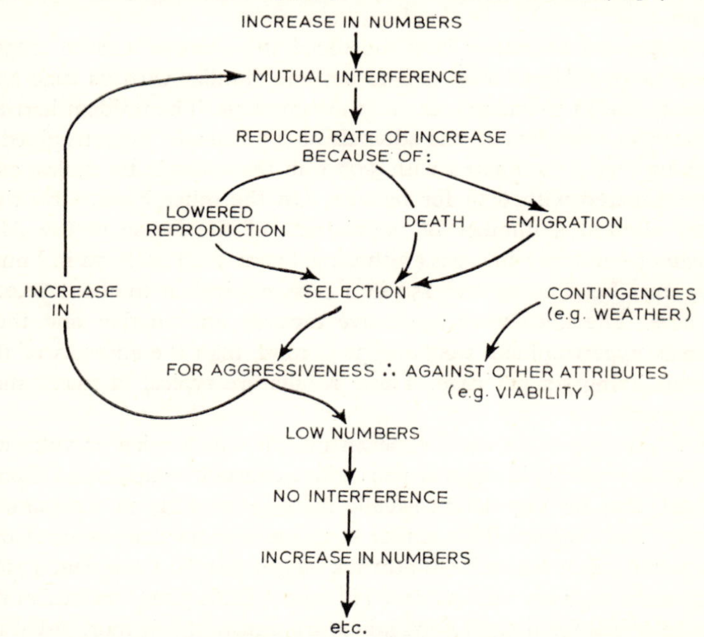

Fig. 5.05. A flow diagram to illustrate Chitty's hypothesis. For explanation see text.

snowshoe hare *Lepus americanus* in which a decline in a natural population had been explained in terms of 'shock disease', a general name for certain physiological derangements typified by low levels of sugar in the blood and glycogen in the liver. Chitty's hypothesis, at this stage, was also influenced by a paper by Christian (1950) in which he suggested that shock disease might be caused in mice by aggression when large numbers of mice were crowded together in a small cage. It is now thought that shock disease is merely an artefact caused by caging a wild animal (Chitty 1959, 1964).

But the original discovery by Christian that social stress can cause hypertrophy of the adrenals and changes in the spleen and thymus has since been confirmed and elaborated many times (Christian 1955a, 1955b, 1956, 1957, 1959, 1961; Christian and Lemunyan, 1957; Christian and Davis, 1964). Christian (1957) proposed the hypothesis that the physiological changes induced by social stress were the cause of the characteristic declines in the numbers of animals in natural populations of microtine rodents that Chitty's hypothesis is intended to explain.

It seems not to matter how the social stress arises. Clarke (1953) allowed a vole *Microtus agrestis* to become familiar with its cage and then introduced a strange vole into the same cage. The resident harried the stranger severely and the adrenals of the stranger hypertrophied – weighing 7·46 g. compared with 5·07 g in the controls for males, and 8·40 compared with 6·59 for females. On the other hand, Christian (1956) allowed a number of populations of the house mouse *Mus musculus* to multiply in cages (with three floors, each 29 × 72 in.) until there were between 90 and 150 in a cage, equivalent to about 80,000 per acre. The mice were aggressive towards one another and their adrenals hypertrophied, weighing 30% more than the adrenals of the controls living six per cage. These results are typical of many such experiments.

It is also typical of such experiments (in which mice or voles are allowed to multiply in cages or pens with a constant or unlimited supply of food) that the population usually increases steadily to a maximum density, and then stabilizes, either at this maximum density, or at one that is not much below it (Southwick 1955a, 1955b; Christian, 1956). Despite the high densities achieved in cages (Table 5.01) 'crashes' of the sort described for natural populations are unusual; it is more usual for a population that has been stable at a high density (say 20–80 × 10^3 per acre) to begin to increase again, perhaps after as much as seven months of stability. Petrusewicz (1957) observed 51 such increases in 49 populations that were maintained for periods varying from one to two years. These results have been repeated consistently leaving little room to doubt that crowding in artificial cages or pens causes: (*a*) increased fighting and social stress, (*b*) hypertrophy of the adrenals with characteristic changes in the relative prominence and histology of different zones of the adrenal cortex, and (*c*) the limitation of population growth associated with a reduced reproductive rate and a characteristic high death-rate among sucklings.

But it seems that there is much room to doubt whether these facts, well established though they are, have any relevance for the ecological facts that Chitty's hypothesis is seeking to explain. The doubt arises because:

(*a*) The densities observed in natural populations are of different orders of magnitude from the densities that are necessary in cages to produce the results that have been discussed above (Table 5.01). With differences of 2–5 orders of magnitude in the densities observed in the two series of experiments it is difficult to imagine that the experiments with animals in cages are measuring the same interactions

Table 5.01. *Densities of populations of microtine rodents at their 'peak' in natural populations compared with densities achieved in artificial pens and cages*

Environment	Species	Density: animals/acre $\times 10^3$	Author
Cage or pen	*M. musculus* (albino)	36–1052	Petrusewicz (1957)
	M. musculus (wild)	4·36–37·7	Southwick (1955a, b)
	M. musculus (wild)	80	Christian (1956)
	Microtus arvalis	30	van Wijngaarden (1960)
Natural	*Microtus agrestis*	0·3	Chitty (1952)
	Clethrionomys glareolus	0·05–0·09	Newson (1963)
	Dicrostonyx groenlandicus	0·05	Pitelka (1958)
	Microtus californicus	0·3–0·5	Krebs (1966)
	Lemmus trimucronatus	0·03	Krebs (1964)
	Microtus spp.	0·03–1·0	Aumann and Emlen (1965)

that occur in natural populations. Indeed the empirical evidence suggests that they are not.

(*b*) Artificial populations in cages and pens usually stabilize largely through changes in the reproductive-rate and failure to recruit young adults, but natural populations usually 'crash' largely through changes in the death-rate especially among juveniles and young adults. The disability in the natural populations usually persists through several generations; but the disability in a caged population can be lifted immediately, in the same generation, by merely transferring the population to an unfamiliar cage. Petrusewicz (1957) allowed 54 populations of white mice to breed in cages until they had stabilized or seemed about to stabilize. Then he changed them to a different cage, a smaller one, a larger one or one of the same size with different furniture. On 36 out of 54 occasions the population began to increase within six weeks and a further seven populations began increasing within 10 weeks. The size of the cage and consequently the direction of change in density seemed to make little difference (Table 5.02).

(*c*) A number of workers have examined the adrenals, thymus and spleen from voles and lemmings in natural populations without finding the differences associated with phases in the growth of the population that are predicted by Christian's hypothesis (Chitty, 1961; Krebs, 1964 and Muller (1965) in Krebs, 1966). But Christian and Davis (1964) criticized some of the earlier conclusions on the grounds that sampling methods did not sufficiently allow for differences associated with age

Table 5.02. *Growth induced in stabilized populations of white mice by changing them to a new cage. Small cage 15 cm × 38 cm, medium 80 cm × 80 cm, large 80 cm × 160 cm.*

Direction of change	Number of experiments	Number of populations with no growth	Increase in density as % of density at time of change
Small to large	20	4	185
Same size	21	4	196
Large to small	13	3	179

and reproductive condition. Chitty's (1967) hypothesis in its modern form no longer leans on Christian's hypothesis. On the contrary Chitty (1960) wrote: 'Indeed there is no difficulty at all in interfering experimentally with reproductive processes; the difficulty comes in producing an effect that corresponds to anything that is going on in nature.'

It remains to consider how well Chitty's hypothesis corresponds to what is going on in nature. In 1952 certain features in the fluctuations in the numbers of voles seemed to indicate a distinctive ecology in these animals. The fluctuations seemed regular enough to be called 'cycles'; declines seemed more likely to occur during summer when the voles should normally be increasing; the adults that were present before the decline set in were heavier than usual; the adults in these dense populations seemed to be no less viable than those in increasing populations; the immediate cause for the decline often occurred when the environment seemed favourable. With increasing knowledge it was found that the course of the cycle was more variable than had been thought; but in 1960 Chitty still thought that the decline that occurred in the population cycles of voles and other microtine rodents was a distinctive phenomenon that called for a special explanation: 'No such solution [i.e. a solution based on the conventional components of environment excluding "aggressors"] can be regarded as satisfactory when there are enough details to show that the phenomenon is no mere reduction in numbers but an association of fairly specific effects that are unlikely to follow except from fairly specific antecedents.' Chitty (1960) gave examples of a number of the 'details' which he thought were important and which I have paraphrased: (1) Declines occur even though the environment seems to be favourable. (2) High density is not sufficient to start an immediate decline. (3) Low density is not sufficient to halt a decline. (4) The vast majority of animals die from unknown causes. (5) The death-rate can be greatly reduced by isolating animals in captivity. (6) The adult death-rate is not abnormally high during years of abundance.

In 1964 Krebs mentioned two qualities in the population cycles of microtine rodents which he said made them distinctive and justified seeking a special explanation for them: 'a cycle is typically a three to four year fluctuation in numbers in microtine rodents characterized by high body weights of adults in the peak summer'. In 1966 Krebs studied eight populations of *Microtus californicus* near Berkeley and found that in some of them the fluctuations were not so cyclical. In some the cycles were like those found in *M. agrestis* by Chitty, but in some the increase and the peak were compressed into one year, and in one the increase, the peak and the decline were all compressed into one year. In other words, as the ecology of the microtine rodents has come to be more widely studied the difference between their ecology and that of other sorts of animals has become blurred.

So, in 1967 the hypothesis which, in its original form, had been advanced on the grounds that 'voles are different' was made general enough to embrace all sorts of animals. Chitty (1967) wrote:

> Most of the evidence for the present hypothesis comes from tests . . . carried out on small mammals; but really crucial evidence is lacking, since neither individual behaviour nor population genetics has so far received much attention from those working on mammal populations. It is thus fortunate that the assumptions on which the hypothesis depends are completely general; hence, contradictory evidence, regardless of the species from which it is obtained should enable one to test the present

interpretation; it cannot both be true for the relevant populations of small mammals and false for those of other groups. . . . An attempt to enlist the help of entomologists has already been made (Chitty, 1965).

I agree with Chitty that the ecology of voles is not likely to be fundamentally different from the ecology of other sorts of animals. Indeed from my experience of fluctuations in the numbers of insects, e.g. *Thrips imaginis* (Davidson and Andrewartha, 1948) or *Austroicetes cruciata* (Andrewartha, 1944), and of opportunistic birds, e.g. *Anas gibberifrons* (Frith, 1967), I would think that Chitty's (1960) points (2), (3), (5) and (6) might refer to many sorts of animals in addition to microtine rodents; and points (1) and (4) are merely statements of ignorance which might be made about any species. With reference to Krebs's (1964) points, too much should not be made of cycles. It would be difficult to maintain with rigour that the fluctuations in the numbers of any microtine rodent are significantly different from random though it may be obvious that they are less erratic than say the fluctuations in a population of *T. imaginis*; but these are better regarded as quantitative not qualitative differences. Neither should too much be made of the healthy condition of the individuals in a population of any sort of animal that has recently been expanding rapidly. The converse might arouse more wonder.

The facts that the numbers of microtine rodents often decline when the environment appears to be favourable and that most of them die from unknown causes may reflect our ignorance of what a vole or a lemming needs to eat. The method for measuring the quantity of food that was used by Newsome (1967) for *Mus musculus* and by Chitty *et al.* (1968) for *Microtus agrestis* has the advantage that it measures what the mouse recognizes as food; what a man judges to be food for a mouse might be quite misleading. Unfortunately the population of *M. agrestis* that Chitty *et al.* measured did not decline as they had expected it to do; so the results of this experiment were inconclusive. I do not know of any occasion when the food of a 'cyclical' rodent was measured in this way during a decline.

An interesting paper by Aumann and Emlen (1965) may indicate a useful direction for research into the food of these small mammals. Aumann and Emlen collated, from 55 publications, information about 118 populations of *Microtus* spp. which, they said, were reported to be 'cyclical'. They inferred the sodium content of the soils in the areas where the rodents lived, either from published soil surveys, or from publications about the geography or the botany of the area. The results of their investigation are given in Table 5.03.

Table 5.03. *The relationship between the peak densities of populations of Microtus spp. and the sodium content of thes oil in the place where they live (After Aumann and Emlen, 1965)*

Sodium content of soil	Number of populations	Maximum density for any population animals/acre	Populations with peak >30	
			Number	%
Low	63	230	12	16
Medium	22	400	16	73
High	33	>1000	30	91

These figures are provocative. Whether they imply that populations 'crash' because they run short of some salt-rich component in their diet, or whether there is another explanation remains to be seen.

If it turns out that microtine rodents 'crash' for the quite conventional reason that they run out of some essential component of their diet the idea of 'aggressors' will become redundant; and, in keeping with the principle of parsimony, should be dropped. In the meantime Chitty's hypothesis holds the field and I retain, tentatively (but I think temporarily), the idea of 'aggressors' in my theory of environment. Of course the fact that voles are aggressive towards their own kind would still require an explanation, but the explanation might be sought in relation to resources (section 3.21) or mates (section 4.3).

Andrewartha (1959) suggested that Chitty's hypothesis might explain the extraordinarily persistent fluctuations in 'caged' populations of flour beetles *Tribolium castaneum*. In Park's experiments the beetles lived in vials in eight grams of flour that was renewed every 30 days; they were kept at 29·5°C in a relative humidity varying between 60–70%. With flour beetles aggression takes the form of eating each other's eggs and juveniles (pupae). I postulated that natural selection for aggression might proceed in opposite directions in dense and sparse populations. After spending about four years in fruitless search for the sort of genetical changes that the hypothesis predicted I abandoned it. A few years later Niven (1967) simulated the curve shown in Fig. 5.06. She wrote into her computer programme empirical information about the life cycle, the life-table and the cannibalism of *Tribolium*, but there was no need to include any genetical factor in order to simulate the natural fluctuations.

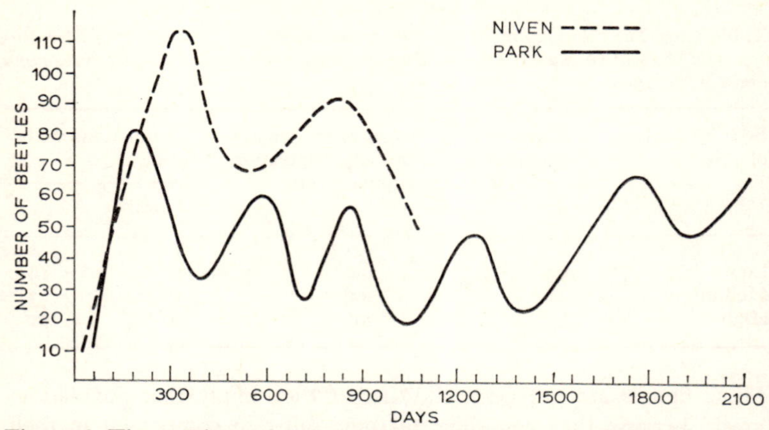

Fig. 5.06. The numbers of *Tribolium castaneum* counted by Park at intervals of 30 days for six years; the numbers in an artificial population simulated on a computer by Niven. The computer calculated the number for every day but recorded them every 30 days.

6 *Components of environment: weather*

6.0 INTRODUCTION

In ecology weather has a technical meaning which differs in certain important respects from common usage. The technical meaning is more limited than the colloquial one because it is convenient to consider under this heading only those stimuli, e.g. temperature, moisture, photoperiod, that evoke physiological responses from the animal. Certain elements of weather, familiar in the colloquial usage, are excluded from the technical meaning because they are better considered in relation to other components of environment. For example, light, measured as photoperiod, must be included because animals have internal 'clocks' which measure the length of day (section 6.3); but light measured in units of energy does not directly influence the physiology of an animal, though it may be an important resource for an autotroph. Similarly wind does not evoke a direct physiological response in an animal, but many animals have adaptations that allow them to be dispersed by turbulence and currents in air or water (section 10.41) just as others have adaptations that allow them to be dispersed by clinging to a moving object (section 10.43). Such aids to dispersal, the moving current or the moving object, are best thought of as resources (section 3.0).

In this chapter I discuss the physiological responses of animals to temperature, moisture and photoperiod, and adaptations in physiology and behaviour that may enhance an animal's chance to survive and reproduce when it is exposed to the range of temperature, moisture or photoperiod that it might experience in the places where it usually lives. These are the important constituents of weather as a component of environment. In the context of ecology weather refers to measurements taken in the precise place where the animal is living.

The influence of weather on an animal's chance to survive and reproduce may be described by a generalized curve like the one in Fig. 6.01.

There is a tolerable range which permits life; the animal's chance to survive and reproduce increases to a maximum within the tolerable range and declines to zero towards the extremes of the range.

Because the body temperature of a poikilotherm depends on the temperature of its environment, because poikilotherms are small and

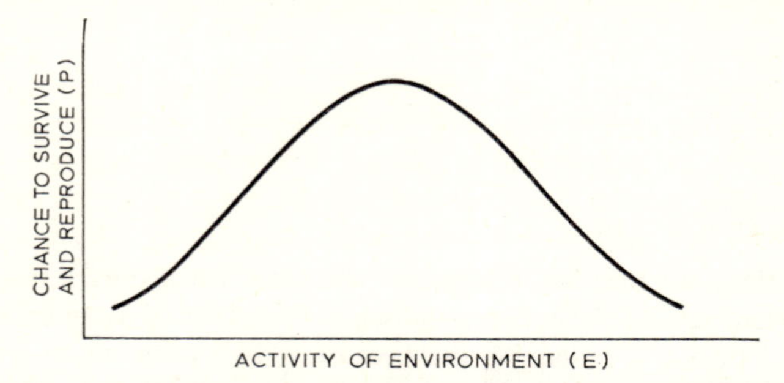

Fig. 6.01. The animal's chance to survive and reproduce first increases and then decreases as the activity of the weather increases from the unfavourable low range through the optimum and into the unfavourable high range. *E* stands for the activity of the environment (weather); *P* stands for the animal's chance to survive and reproduce.

because temperature and humidity are often interdependent, temperature and humidity are more often important in the environments of poikilotherms than of homoiotherms. Of course, this distinction does not hold when weather operates further out on the ecological web, influencing the animal's chance to survive and multiply indirectly, by modifying the activity of some other component of environment, e.g. food.

6.1 TEMPERATURE

Temperature influences the speed of development, the duration of life, the fecundity, and the behaviour of animals, especially poikilotherms.

6.11 *The influence of temperature on speed of development*

At 25°C *Thrips imaginis* completes its development from egg to adult in about nine days and the weevil *Calandra oryzae* requires about 35 days at the same temperature. A convention has grown up among zoologists to use the reciprocals of such measurements as a measure of the speed of development of the animal. Thus it is said that the speed of development of *Calandra* at 25°C is 2·9% per day; but the speed of development of *Thrips* at 25°C is 11·1% per day. It is immediately clear that these two measures of 'speed' are not to be compared with each other because they refer to proportions of quite different quantities. The usefulness of this method is limited to comparing the speed of the same developmental process (i.e. the same stage in the life-cycle of members of the same species) at different temperatures. It has been widely used for this purpose.

In nature, animals usually live in fluctuating temperatures. But it is difficult to do experiments with fluctuating temperatures; also the results of such experiments are difficult to interpret. Andrewartha and Birch (1954, section 6.232) reviewed the literature on fluctuating temperatures and concluded that the principles relating to constant temperatures were broadly relevant to fluctuating temperatures subject to certain simple provisos which applied in particular circumstances.

6.111 The speed of development at constant temperatures

Figure 6.02 gives the results of Powsner's (1935) experiments with the eggs of *Drosophila melanogaster*. The points on the descending curve represent the raw data as it came from the experiment; they show the number of hours that the eggs took to complete their development at each temperature. The points on the ascending curve are the reciprocals of the original data and represent the conventional measure of the speed of development.

The relationship between temperature and speed of development has been described by the particular sigmoid curve that is called the logistic curve. The general formula for the logistic curve is:

$$\frac{1}{Y} = \frac{K}{1 + e^{a+bt}}$$

where Y is the duration of development at temperature t, and K, a and b are empirical constants. For an ascending curve the constant b is negative.

In Fig. 6.02 the empirical points lie closely along the calculated curve up to 30°C. Above 30°C the harmful influence of high temperature causes the observed points to fall below the theoretical curve. If the experiment had been continued at still higher temperatures the empirical points would have approached the base line as in Fig. 6.01. Between the limits 17 to 25°C the curve is rather flat and a straight line would fit the points in this range fairly well.

Until Davidson (1944) pointed out the superior merits of the logistic curve the straight line had been widely used. The straight line is the reciprocal of the hyperbola. The general formula for the hyperbola is:

$$Y = \frac{K}{t - d}$$

where Y is the duration of development at temperature t, and K and d are empirical constants.

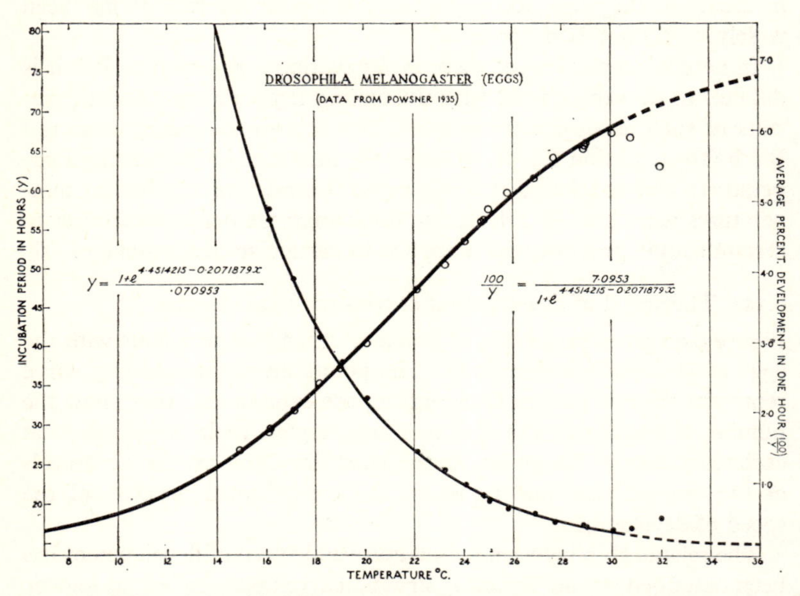

Fig. 6.02. The relationship between 'speed of development' and temperature of the eggs of *Drosophila melanogaster*. The hours required to complete the egg-stage are plotted against temperature (*descending curve*) and the reciprocal of this time is plotted against temperature (*ascending curve*). The points plotted on both curves are observed points. The ascending curve is a logistic curve and the descending one is its reciprocal, as indicated by the formulae shown on the diagram. (After Davidson, 1944.)

For the reciprocal of Y, which is the conventional measure for speed of development, this equation can be written:

$$\frac{1}{Y} = k + bt$$

where Y and t have the same meanings and k is a new constant equal to $-bd$. This is the equation of a straight line and d can easily be shown to be the value of t when $1/Y$ becomes zero. It has been customary, therefore, to call d the 'threshold of development' and (this leading to the error that Weiss complained about – section 1.3) to imagine that all development ceases at the so-called 'threshold of development'. But it is now well known, because any well-conducted experiment will demonstrate it, that the temperature-time curve approximates to the hyperbola for only a very short range of temperature; and the concept of *d* as the 'threshold of development' is unrealistic.

6.12 *The lethal influence of temperature*

The lethal influence of low temperature has been the subject of much study, especially by entomologists in Europe and North America because they have been interested in the survival-rates during winter of various insect pests. The lethal influence of high temperature has received much less attention. With terrestrial animals, especially poikilotherms, the difficulty may be to measure the influence of heat independently of dryness. In nature harmful high temperatures are more likely to occur in deserts or in warm temperate zones where the summer is arid. In these circumstances dryness may be more important than heat. High temperature may be more important for certain species of fish that live in shallow lakes or rivers in temperate climates. There have been a number of experiments done on fish to determine the range of temperature within which they can live and the influence of acclimatization on this range. High temperature may also be a risk for animals that live in well-insulated places. Birch (1946) found that a dense population of grain weevils *Calandra oryzae* living in large bins of wheat might, by their metabolism, cause the temperature of the wheat to rise until all the weevils died except those living near the outside.

The influence of lethal temperature cannot be measured by the same direct methods that are appropriate for measuring the influence of temperature on the speed of development. For example one of Johnson's (1940) experiments with the eggs of *Cimex* which I describe below was designed to find out how long the eggs would live at 4·2°C. While the eggs remained at 4·2°C there was no way of telling by inspection which were dead and which were still alive. In order to discover this they had to be removed from the cold and incubated in the warmth. Then it was possible to tell what proportion of the sample had died during exposure to cold but it was not possible to tell for any one egg what 'dose' of cold had been required to kill it. In other words, there was no way of measuring directly the susceptibility of an egg to cold and consequently no direct method of estimating the mean susceptibility of the eggs to cold as we could, for example, estimate their mean weight or the mean duration of their life-cycle. The way around this difficulty is to design an experiment that allows us to estimate the dose that is required to kill just half of the population. This provides an estimate of the median which, for a normal curve, is equivalent to the mean (section 11.2).

If the problem is to measure the duration of exposure to a particular temperature that is required to kill half the population, we begin by exposing a number, say six samples of 50 or more animals each, to this temperature for varying periods. If the problem is to measure the temperature that is required to kill half the population when they are exposed to it for a particular period, then all the samples are exposed for the same period but to different temperatures. In either case the treatments are chosen with the expectation that the least severe will kill a small proportion of the sample, the most severe will kill about 80% and the remainder will straddle the 50% mark as nicely as may be. It may be necessary to keep one of the samples as a control, i.e. it is not exposed to the harmful temperature at all. As each sample completes its treatment it is removed to a favourable temperature, kept for a suitable period, several days or more, and the number of deaths recorded.

If these numbers, expressed as proportions, are plotted against the treatment (or dose) they will fall along a sigmoid curve, which is the familiar 'dose-mortality' curve of the toxicologist. This curve is also the one that can be got by progressively summing the ordinates of the ordinary frequency-polygon. Provided that the frequency-distribution of deaths against 'dose' is normal or nearly so, the sigmoid 'dose-mortality' curve becomes a straight line when the proportions are transformed to probits (Finney, 1947). If the distribution is not normal it may be made so by transforming the dose to logarithms or some other appropriate scale.

Once the straight line has been calculated, it is easy to read off, by linear interpolation, the dose (temperature or duration of exposure) that would be expected to kill any proportion of the population. The dose required to kill 50% corresponds to the mean dose that we could have calculated directly if we were able to write down by direct inspection of each animal the dose just required to kill it. We may sometimes require to know the dose required to kill some other proportion of the population. There is no difficulty in this, but it must be remembered that, unless the size of the sample is increased, the estimates will be less precise. To estimate the dose required to kill 95% the sample requires to be three times, and for 99% 10 times, as large as that required to estimate, with the same degree of accuracy, the dose required to kill 50%. It is usual, in writing about probit analysis, to abbreviate 'the lethal dose required to kill just 50 per cent' into 'L.D. 50'.

Johnson (1940) used the probit method to measure the influence of low temperature on the survival-rate in eggs of the bed bug *Cimex*. This paper should be studied in the original by the student who is interested in technique.

In one experiment, eggs of a uniform age (0–24 hr at 23°C) were exposed to three temperatures, 4·2°, 9·8° and 11·7°C for periods ranging from 9 to 50 days. At the conclusion of each exposure the eggs were incubated at 23°C. Those that did not hatch were considered to be dead. The results are set out in Table 6.01. The death-rate is expressed both as a per cent and as a probit. The former is transformed to the latter simply by reference to the appropriate Table in Finney (1947) or Fisher and Yates (1948). Table 6.01 also gives the values for b, L.D. 50 and L.D. 99·99. The first is the coefficient for linear regression of probit on dose; it measures the slope of the probit line and inversely the variance.

Table 6.01. *Death-rate of eggs of Cimex exposed for various periods to 4·2°, 9·8° and 11·7°C**

Exposed to 4·2°C			Exposed to 9·8°C			Exposed to 11·7°C		
Exposure (days)	Death-rate %	Death-rate Probit	Exposure (days)	Death-rate %	Death-rate Probit	Exposure (days)	Death-rate %	Death-rate Probit
9	4·2	3·27	7	10·1	3·72	7	9·4	3·7
16	16·9	4·04	14	7·5	3·56	14	9·7	3·7
23	52·5	5·06	21	19·9	4·15	21	36·7	4·7
30	86·3	6·09	28	65·8	5·41	29	59·2	5·2
35	94·5	6·60	35	86·3	6·09	35	69·4	5·5
40	91·8	6·39	42	100·0	8·72	42	90·7	6·3
44	94·5	6·60				46	94·0	6·6
50	100·0	8·72				49	96·2	6·8
b	0·1139 (0·000151†)			0·1273 (0·000241)			0·0800 (0·000058)	
L.D. 50	22·9 days (1·120)			25·8 days (0·774)			26·7 days (1·409)	
L.D. 99·99	55·6 days			55·0 days			73·3 days	

* After Johnson (1940).
† Figures in parenthesis are variances.

The L.D. 50 and L.D. 99·99 are the estimated durations of the exposures required to kill 50% and 99·99% respectively. The variances for *b* and L.D. 50 are also included because these quantities are necessary for the calculation of the statistical significance of the results.

Figure 6.03 A shows the death-rate, in %, plotted against the duration of exposure, and Fig. 6.03 B shows the same data plotted as probits (see also Figs. 11.21*a*, 11.22*a*). The advantages of transforming the data to a scale in which the regresssion becomes linear are that it enables us to make a precise estimate of the exposure required to kill any proportion of the population; and an exact significance can be attributed to the observed differences. Figure 6.03 B indicates that the probit lines for 4·2° and 9·8°C are nearly parallel and that both slope more steeply than the probit line for 11·7°C. On the other hand, if we compare the co-ordinates of the abscissae for probit 5 (i.e. the value for L.D. 50) we find that 9·8° and 11·7°C come close together whereas 4·2°C is well away to the left. These two statistics, *b* and L.D. 50, are the two most instructive ones to extract from the probit line. The L.D. 50 is the most reliable measure of the animal's capacity to withstand exposure to a given temperature, and *b* is the best comprehensive measure of the relative toxicity of short and long exposures. A large value for *b*, i.e. a steep probit line, indicates that there is little 'scatter' between the exposures that kill few and many; conversely a small value for *b*, i.e. a gently sloping probit line, indicates a wide margin between the exposures required to produce these extremes in the death-rate (section 11.2).

Kozhantchikov (1938) recognized that insects could be divided into three groups according to their response to low temperature. (1) The ones in the first group cannot survive for any considerable period if the temperature falls below the lower limit of the range favourable for normal development. They cannot become dormant at low temperatures; they must either develop or die. This group is made up of (*a*) species that live in or originated from tropical or sub-tropical climates, e.g. *Cimex*, *Calandra* and *Locusta*; there are many of these species in which no stage in the life-cycle is capable of becoming dormant if the temperature falls too low for development; (*b*) species from temperate climates in which there is a certain stage in the life-cycle that is adapted to survive the winter, but all other stages resemble the tropical forms in lacking the capacity to become dormant at moderate or low temperatures. (2) The second group in Kozhantchikov's classification comprises the forms that can become 'quiescent' as Shelford (1927) defined this term. That is, while remaining competent, at any time, to develop if they are placed in the warmth, they may nevertheless survive, healthy but inactive, at temperatures too low for develop-

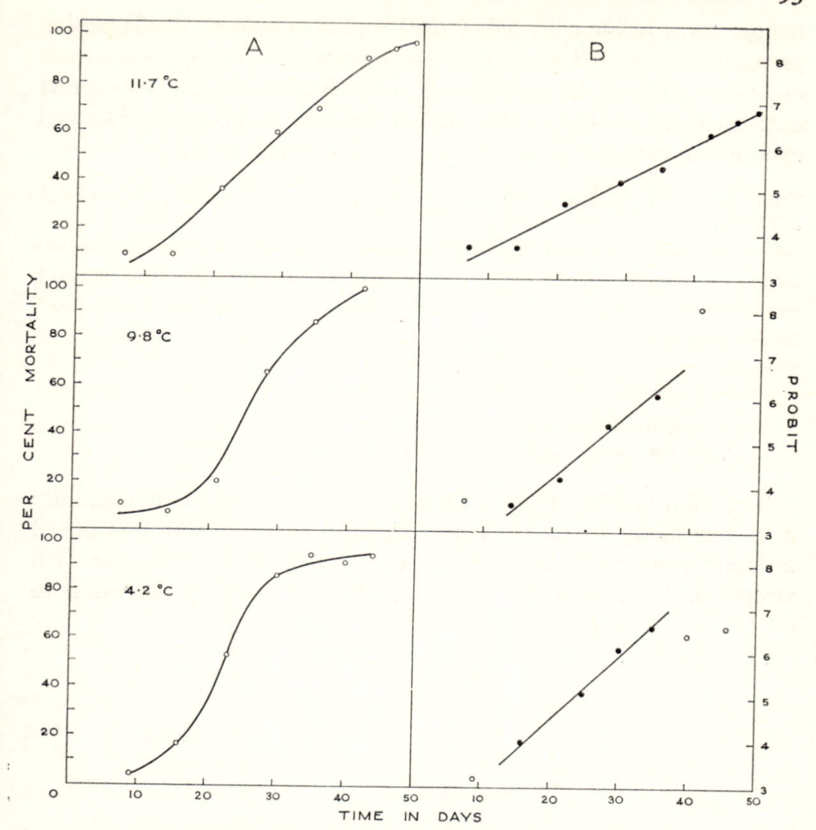

Fig. 6.03. The death-rate for eggs of *Cimex* exposed to 4·2, 9·8 or 11·7°C for a varying number of days. A, death-rate expressed as a per cent is plotted against time; B, by transforming from per cent to probit the sigmoid curve becomes a straight line. (Data from Johnson, 1940.)

ment. Contrary to popular belief this group is very small; in searching the literature it is difficult to find an authentic representative of it. (3) The third group contains the special over-wintering stage of the species adapted to the cold-temperate climates. Cold-hardiness and dormancy is nearly always associated with a condition which is known as diapause (Andrewartha, 1952).

The sawfly *Cephus cinctus* is indigenous to the prairies of western Canada where it has become a pest of wheat. It spends the winter as a prepupa inside the stem of the wheat where it may be exposed to temperatures below 0°C. Salt (1950) kept a sample of 18 diapausing prepupae at −20°C for 120 days; he recorded the day on which each one became frozen. The results are given in Table 6.02. It seems that cold-hardy insects like *Cephus* are not likely to be killed by prolonged exposure to cold unless they freeze, and not always then.

Table 6.02. *Time in days required for freezing to occur in overwintering prepupae of Cephus**

Day	1	2	3	4	5	8	13	15	16	28	33	42	66	71†
No. of prepupae freezing	0	0	2	1	1	1	1	1	1	1	1	2	1	1

* Data from Salt (1950)
† On the 120th day there were still four prepupae unfrozen.

The solid line in Fig. 6.04 illustrates the way that the temperature of a non-hardy insect (*Ephestia*) changed when it was transferred abruptly from room temperature to −30°C. The temperature fell steeply and smoothly to −24·3°C, when it abruptly increased to

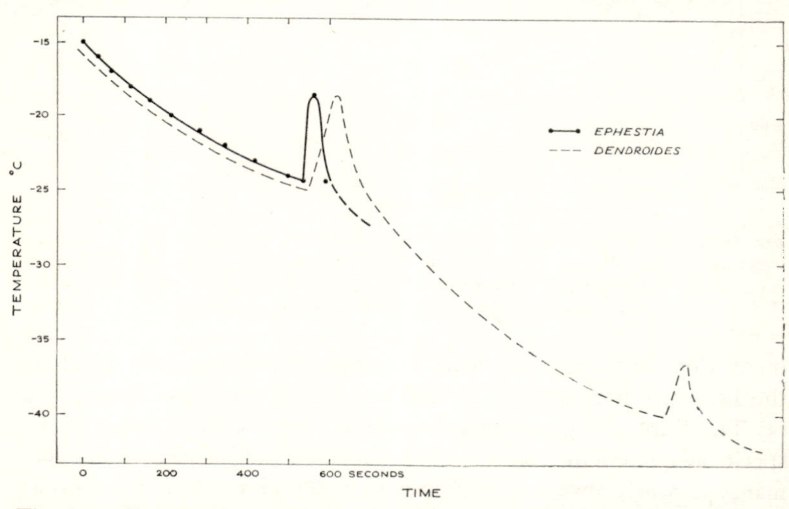

Fig. 6.04. Characteristic curves for the temperature of an insect that has been abruptly transferred from room temperature to some low temperature, say −40°C. The complete line is characteristic of most species which die after the first rebound. The broken line is characteristic of the few very hardy species that can survive the first rebound. (After Andrewartha and Birch, 1954.)

−18·6°C and then proceeded to fall again; the wave of crystallization that followed the freezing at −24·3°C liberated enough heat to raise the temperature by 5·7°C. Then, as this heat was absorbed the temperature began to fall again. The temperature at which cooling ceased is called the 'undercooling point'; the rise in temperature is the 'rebound'; and the temperature to which the rebound rises is called the 'freezing point'.

6.13 *The limits of the tolerable zone*

If an animal dies promptly when it is exposed to an extremely high or low temperature there may be no difficulty in attributing its death to the exposure to extremes of temperature. At less extreme temperature

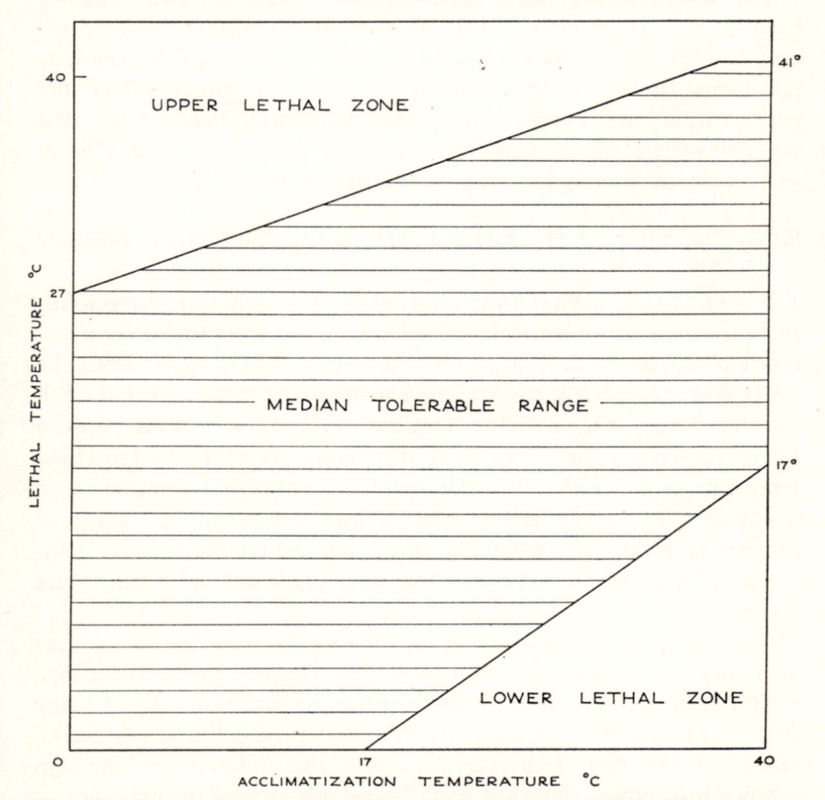

Fig. 6.05. The median tolerable range of temperature for goldfish. Notice how the limits of the tolerable range may be influenced by the temperature at which the fish had been living. (After Andrewartha and Birch, 1954.)

death may take longer but it may still be recognized as being largely caused by exposure to unfavourable temperatures. But as we move farther and farther away from the extremes we come to a zone of moderate temperature where it is not reasonable to attribute any lethal influence to temperature. Fry (1947) called the two limits to this median tolerable zone the 'incipient lethal low temperature' and the 'incipient lethal high temperature' and he defined the extremes outside these limits as the 'upper lethal zone' and the 'lower lethal zone' respectively. Can we discover the limits of the median tolerable zone by experiment?

When we come to plan the experiment, we find that it cannot be done precisely because, as the temperature becomes less extreme, the exposure becomes so long that we may not be able to decide whether the animal died from old age or from some harmful influence of temperature. In their experiments with young trout, Fry *et al.* (1946) used periods that varied from 20 to 80 hours depending on the circumstances: the periods were long enough to justify the assumption that any fish that was still alive at the end of the experiment would have been able to survive an indefinitely long exposure to that particular temperature.

6.131 The influence of acclimatization on the limits of the tolerable zone

Fry *et al.* (1942) measured the limits of the tolerable zone for goldfish that had previously been acclimatized to temperatures ranging from just above 0° up to 40°C. The results are shown in Fig. 6.05. Both the lower and the upper incipient lethal temperature were higher for fish that had been living at higher temperature; and the distance between the two limits was narrower for fish that had been acclimatized to a high temperature. The incipient lethal high temperature could not be raised above 41°C by acclimatization. Taking into account the full influence of acclimatization the lowest incipient lethal low temperature for goldfish may be about 0°C and the highest incipient lethal high temperature about 40°C.

The ability to extend the limits of the tolerable zone by acclimatization may be valuable to fish that live in shallow water. Huntsman (1946) showed that extremes of temperature sometimes killed many fish in Ontario. Brett (1944) collected samples of bullhead (*Ameiurus*) from Lake Opeongo at intervals during the year and measured the high temperature required to kill half the sample during an exposure of 12 hours. Figure 6.06 shows that the seasonal trends in 'heat-hardiness' were large and striking.

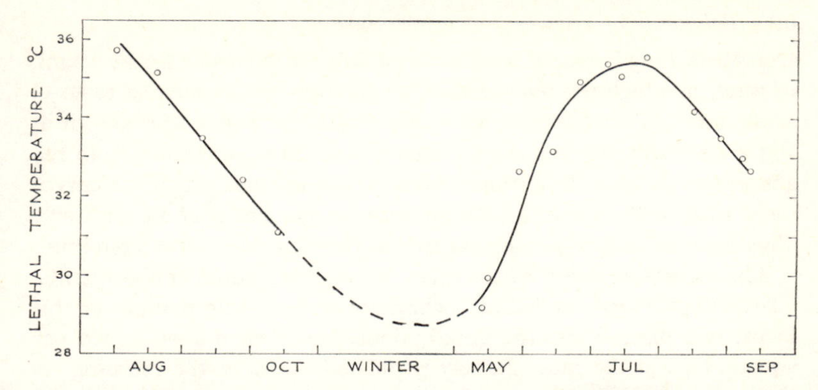

Fig. 6.06. The seasonal trend in the 'heat-resistance' of the bullhead *Ameiurus*. The ordinate gives the L.D. 50 for an exposure of 12 hours to high temperature. The fish clearly have become acclimatized to the prevailing temperature of each season. (After Brett, 1944.)

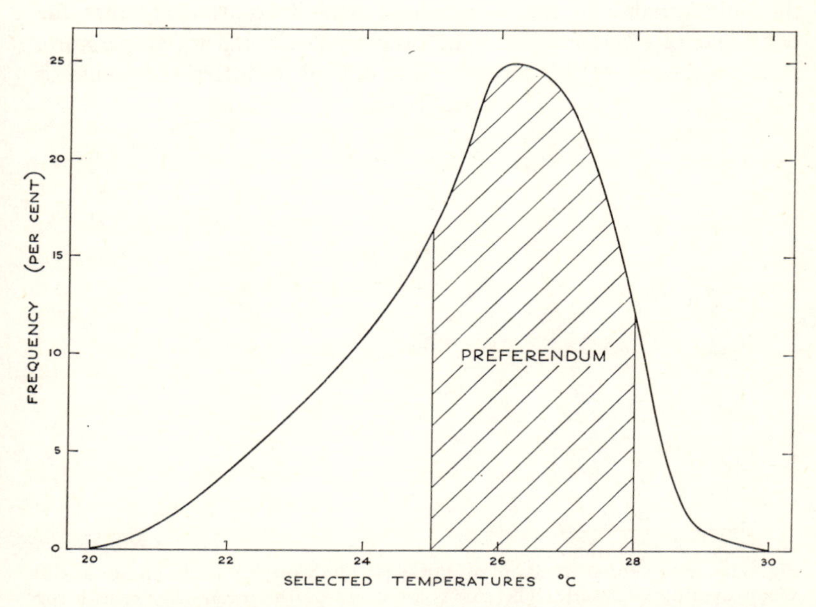

Fig. 6.07. The 'preferendum' for the fish *Gisella nigricans*. The ordinates indicate the per cent of the fish choosing each temperature. About 75% of the fish congregated in the range indicated by the shading. (After Andrewartha and Birch, 1954.)

6.14 *Behaviour in a gradient of temperature*

Doudoroff (1938) placed a number of fish *Gisella nigricans* in a tank of water in which the temperature varied from 20° at one end to 30°C at the other. They arranged themselves with the frequencies shown in Fig. 6.07. Most of the fish congregated near 26°C; ordinates from the abscissa at 25° and 28° enclose about 75% of the area under the curve: more than 60% of the fish stayed between the limits of 25° and 27°. This band of preferred temperature has been called the 'preferendum'.

The locust *Chortoicetes terminifera* has a preferendum at about 42°C. Clark (1947, 1949) studied the behaviour of the young nymphs of this locust in a place where the initial population reached about 2,000 per square yard in the most densely populated parts. In the morning, as the sun warmed the ground, the young locusts followed the gradients in temperature until they came to congregate in the places where the temperature approached more closely to 42°C.

The results of Doudoroff's experiments with *Gisella* may be attributed without any reasonable doubt to temperature because temperature was the only variable in this experiment. And it seems likely that the behaviour of *Chortoicetes* may be properly attributed to temperature. But experiments of this sort are often difficult to interpret because of

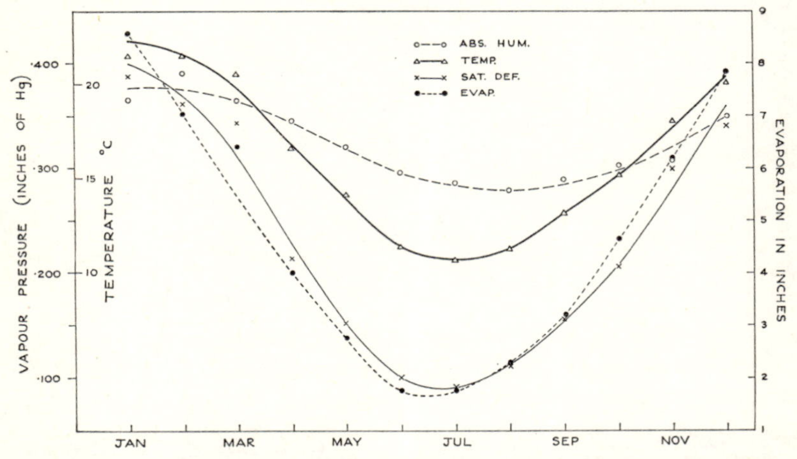

Fig. 6.08. The annual trends in atmospheric temperature and humidity at the Waite Institute, Adelaide. The curves are based on the mean daily records for 23 years. Note that changes in absolute humidity are slight and gradual compared to the more abrupt changes in saturation deficiency and evaporation, which are closely related to temperature and absolute humidity. (After Andrewartha and Birch, 1954.)

interactions between temperature and other components of weather. In water the concentration of dissolved oxygen is likely to vary with temperature. In air changes in absolute humidity are slow and small relative to temperature; consequently, fluctuations in the evaporative power of the atmosphere, whether measured directly as 'evaporation' or indirectly as 'saturation deficit', follow fluctuations in temperature (Fig. 6.08).

6.15 *Adaptations to temperature*

'Acclimatization' refers to physiological changes that occur in an individual during its lifetime; they are usually reversible. 'Adaptation' refers to genetical differences that have arisen by mutation and selection. They are reversible only by the same evolutionary processes.

Moore (1939, 1942) compared the speed of development of the eggs of northern species of frogs (*Rana*) with those of southern species (Table 6.03). He used the time taken to develop from stage 3 to stage 20 as a

Table 6.03*. *The duration of development of the eggs of four species of Rana at three different temperatures in relation to the temperature of the water where they usually breed*

Species of *Rana*	Northern limit of distribution (°latitude N)	Usual temp. of water at breeding time (°C)	Hours required to develop from stage 3 to stage 20			Ratio column (b) to (c)
			12°C (a)	16°C (b)	22°C (c)	
R. sylvatica	67°	10°	205	115	59	1·95
R. pipiens	60°	12°	320	155	72	2·15
R. palustris	51°–55°	15°		170	80	2·12
R. clamitans	50°	25°		200	87	2·29

* After Moore (1942)

criterion. These are arbitrary stages in the morphogenesis of amphibian embryos that were described by Pollister and Moore (1937). There is a close relationship between the temperature of the water where the species usually breeds and the speed of development of the eggs at constant temperatures. Those from cold places developed more rapidly at low temperatures than those from warmer places. But the acceleration with rising temperature was greater for those from warm places; the most southern species *R. clamitans* developed from stage 3 to stage 20 in 45 hours at 33°C – a temperature that killed the other three species.

Similar adaptations to temperature may be observed with respect to the limits of the tolerable range. Brues (1939) found the larvae of certain Diptera living in thermal springs in Indonesia at temperatures up to 52°C. At the other extreme the beetle *Astagobius angustatus* lives in ice-grottos where the temperature ranges between −1·7° and 1°C (Allee *et al.*, 1949). But no individual species is known that can thrive over such a wide range as from 0° to 50°C. The range of temperature

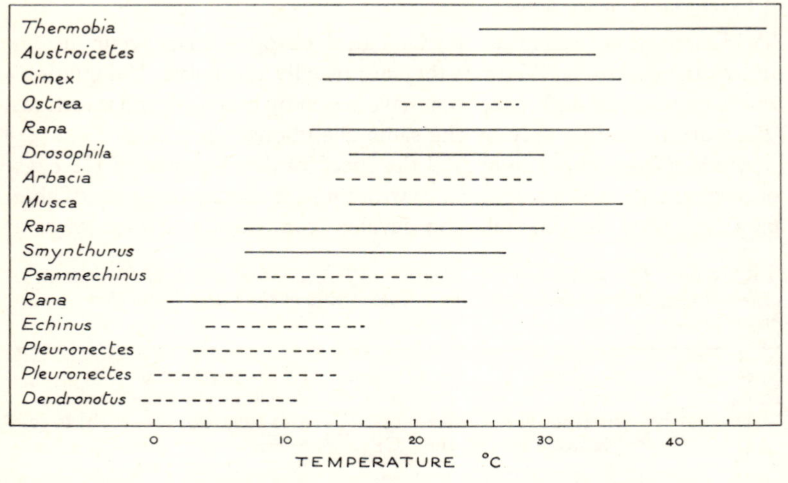

Fig. 6.09. The range of temperature favourable for the development of a number of aquatic (broken lines) and terrestrial or amphibious animals (solid lines). Note (*a*) that the ranges for the aquatic species are usually shorter than for the terrestrial ones and (*b*) the wide range between species in the limits of the favourable range. This is related to the temperatures that prevail in the places where the animals live. The three lines for *Rana* refer to three different species. (After Andrewartha and Birch, 1954.)

favourable to a particular species is related to the prevailing temperatures in the places where the animals usually live. Not only the average or usual temperature but also the range of temperature are reflected in the physiology of the animals. Figure 6.09 shows that for species from cold places, like *Dendronotus*, the favourable range is low and for those from hot places, like *Thermobia*, it is high. Also the limits of the favourable range are wider for terrestrial than for marine species, which reflects the greater variability of temperature on land than in the sea.

6.2 MOISTURE

About the wettest place in which an animal may live is fresh water. Certain 'terrestrial' animals such as rotifers, nematodes and tardigrades live in films of water that surround for example, rotting vegetation; their places of living can hardly be discriminated from those of aquatic animals living in fresh water. About the driest place for an aquatic animal is a rock pool of sea-water that is drying up. But some truly terrestrial animals live in drier places than the 'driest' of the aquatic places. Some aquatic animals have a dormant stage in the life-cycle that can remain alive in very dry terrestrial places. In the next few paragraphs I mention a few of the many different sorts of dry places where animals live in order to illustrate the drought-hardiness that may be found in diverse sorts of animals.

The larvae of *Polypedilum vanderplanki*, a small midge of the Family Chironomidae, live in temporary pools formed in shallow hollows in rocks in Uganda. In their normal active condition water makes up about two-thirds of their total weight. When the pool dries up the larvae dry up also, becoming virtually as dry as the dust in which they are living. They survive the drought in this desiccated condition; when the next lot of rain fills the pool again they replenish their tissues with water and resume development. Hinton (1960) showed that when larvae in this condition of *cryptobiosis* were stored in air of relative humidity 60% the water-content of their bodies was about 8%. Some larvae that had been stored in dry dust in the laboratory for 39 months revived when they were placed in water, and subsequently they completed their development, apparently normally. Cryptobiosis occurs in other invertebrates, not necessarily closely related to the Chironomidae, notably in rotifers, tardigrades and nematodes.

The snail, *Helicella virgata*, seems to seek out places that are especially hot and dry when it is settling down for the summer; the top of a fencing post or the exposed northern face of a stone wall are characteristic places. The snail, having settled in such a position, may spend the entire summer there without food or drink; or it may wake up briefly during a shower of rain only to resume its position when the rain dries up. During the five months that the snail sits in the exposed position it might evaporate about 120 mg of water; a standard evaporimeter tank might evaporate 36 inches of water during the same time.

The kangaroo tick, *Ornithodorus gurneyi*, lives in the dust about half an inch below the surface of kangaroo 'wallows' in country where the annual rainfall is about seven inches and annual evaporation about six feet. The ticks have no opportunity to eat or drink, except when a kangaroo happens to use the wallow. But the ticks can remain alive and conserve the water-content of their bodies for several months without food or drink.

The flour moth, *Ephestia kuhniella*, can grow from an egg, weighing less than 1 mg to a pupa weighing 16 mg (of which 10 mg is water) in flour that has been dried in an oven and kept in a closed container over concentrated sulphuric acid (i.e. in air virtually at zero relative humidity). Even though it is eating such dry food and respiring such dry air the caterpillar can retain sufficient of the water of metabolism to maintain a water content of 64% in its body.

The kangaroo rat, *Dipodomys merriami*, lives in the desert in Arizona, eating only dry grain and not drinking at all. It could (on occasion), if it would, gain water by eating succulent vegetation, but it seems never to do this. The kangaroo rat, like the trap-door spider (section 6.21), avoids the worst rigours of the desert by digging itself a burrow, but, unlike the trap-door spider, this is only part of the explanation of its success (section 6.223).

The camel, being too large to escape the heat by going underground, lives and works on the surface exposed to the full severity of the desert sun. There is an authentic record of a journey of 600 miles made by camels in the Empty Quarter of the Sahara where there is no drinking water (Schmidt–Nielsen, 1964).

Finally, the brine shrimp, *Artemia salina*, lives in pools of sea-water and can thrive in sea-water which is drying up and from which the dissolved salts are crystallizing out of solution (Croghan, 1958a).

Cryptobiosis is an unusual adaptation not found in many sorts of animals. But many insects reduce the water-content of their bodies before entering diapause. For example, the fully grown larva of the moth *Diacrisia*, which overwinters as a diapausing caterpillar, changed in weight from 600 to 200 mg in the autumn. Water accounted for most of the weight that was lost. In the spring it absorbed water and nearly regained its original weight before pupating (Payne, 1927). Certain earthworms are reported to have survived brief periods of desiccation down to 70–80% of their original water (Schmidt, 1918; Hall, 1922); and Thorsen and Svihla (1943) claimed that certain frogs survived brief periods of desiccation down to 38–60% of their original water. The camel is remarkable among mammals because it can sustain its normal activities while losing water equal to 30% of its weight, and it may be desiccated and rehydrated repeatedly as a normal part of its life (Schmidt–Nielsen *et al.*, 1956).

These are important exceptions, but the general rule is that desert-dwellers contain about the same proportion of water in their bodies as related species living in moister places. For example the pupae of *Ephestia kuhniella* that were reared in oven-dry flour contained 64% of water compared with 68% in pupae that had been reared in flour containing 10% of water (Fraenkel and Blewett, 1944). The kangaroo rat, *Dipodomys merriami*, living in the desert in Arizona on air-dry barley, contained 66% of water compared with 67% when it had a moist diet. For comparison an ordinary white rat that was allowed to drink as much water as it would, contained 69% of water (K. Schmidt-Nielsen and B. Schmidt-Nielsen, 1952). When *Dipodomys* were desiccated by being offered nothing but dry soy bean (40% nitrogen) for food and nothing to drink they lost about 34% of their original weight before dying, but at death they still contained 67% water. When *Tribolium confusum* were starved in dry air (0% relative humidity) for seven days they lost weight steadily, losing about 30% of their original weight in seven days; but the proportion of water in their bodies remained constant at about 56% (Roth and Willis, 1951). Few terrestrial animals are as hardy as *Ephestia*, *Tribolium* and *Dipodomys*, but most species, delicate and hardy ones alike, need to maintain a relatively high level of water in their bodies if they are to remain alive.

An animal's capacity for maintaining an appropriate level of water in its body when it is living in a very wet or a very dry place might be judged in several ways. (*a*) Against what pressure can it maintain the water-content of its body, i.e. what *force* can it exert? (*b*) How much *work* is done to conserve water? i.e. how much *energy* is used in water-conservation? (*c*) *Power* is the rate of doing work, and it would be pertinent to ask how much *power* can the animal devote to conserving water? (*d*) More can be achieved by a machine of efficient design than by one with less efficient design, so it is also interesting to inquire what are the *mechanisms* that conserve water? (*e*) What behaviour does the animal have for avoiding extremes of wetness or dryness?

As Ewer (1968) has made quite clear the answer to the last question will depend on the answers to the first four because behaviour evolves along with physiology and anatomy. Size is important in relation to both behaviour and physiology as the following discussion ranging through insects, small rodents and camels makes quite clear.

6.21 *Behaviour in relation to moisture*

Moisture in the environment of a terrestrial animal may be conveniently measured either as relative humidity or evaporation (Fig. 6.08). Wellington (1949) postulated that the behaviour of the caterpillars of *Choristoneura* was more closely related to evaporation than to relative humidity, and that the response to moisture would outweigh the response to temperature. Accordingly he designed an apparatus in which he could control temperature and humidity at the same time, producing a gradient in either or both, and he designed a small evaporimeter that measured the rate of evaporation directly. He placed four batches of larvae of *Choristoneura* in four different sorts of gradient which are described in the legend to Fig. 6.10. In each experiment most of the larvae congregated in a zone where the evaporation was between 0·12–0·14 mm^3/min despite big variations in temperature and absolute humidity between experiments. To some extent the response may have depended on the condition of the larvae (although this was not brought out in this experiment) because, on another occasion, Wellington found that similar larvae, that had been starved in dry air for one hour, tended to congregate in a moister zone than larvae that had not been starved.

Roth and Willis (1951) showed that the behaviour of the flour beetles *Tribolium confusum* and *T. castaneum* in a gradient of moisture depended on the condition of the beetles. The insects were placed in an 'olfactometer' which allowed them to move either into a stream of moist air (nearly saturated with water vapour) or a stream of dry air (near 0% relative humidity). Some beetles were taken from the flour containing 10–12% water where they had been living and tested immediately; others were starved in air at about 0% relative humidity for periods ranging from 1 to 6 days. The results are shown in Fig. 6.11: six days' starvation reversed the normal tendency to congregate in dry air; both species were alike in this but *T. confusum* was more 'hardy' than *T. castaneum* because a longer period of starvation was required to produce the same result (exp. 12.11).

Roth and Willis carried these experiments one step further. Beetles (females of *T. castaneum*) that had been starved in dry air for 4 days (by which time only 7% of them were in the condition to choose dry air) were then starved for a further 60–70 hours in moist air (with a relative humidity in excess of 90%) during which time they continued to lose a little weight (from 73 to 72% of the original weight) but the proportion of water in the body increased slightly, from 60 to 62%.

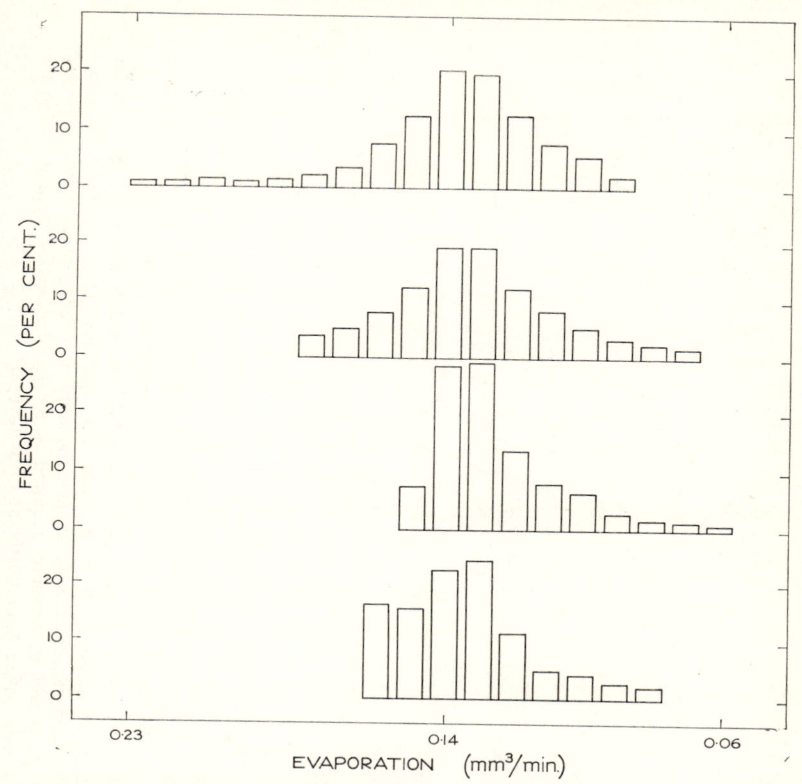

Fig. 6.10. Distribution of second-instar larvae of the spruce budworm *Choristoneura fumiferana* in response to evaporation in four sorts of gradient: A (*top*), in a gradient of temperature in dry air. The isotherm for 36°C lay above the zone where the evaporation was 0·23 mm³/min. B, in a gradient of temperature in moist air. The isotherm for 36°C lay above the zone where the evaporation was 0·18 mm³/min. C, in a gradient of temperature in a very moist atmosphere. The isotherm for 36°C lay above the zone where the evaporation was 0·15 mm³/min. D (*bottom*), in a gradient of evaporation at a constant temperature of 20·6°C. The maximal rate of evaporation did not exceed 0·16 mm³/min. (After Wellington, 1949.)

When these beetles were tested in the olfactometer 42% chose the dry air, i.e. 60–70 hours' starvation in moist air had allowed 35% of the beetles to regain their preference for dry air. Another lot of beetles (males of *T. confusum*) were starved in dry air for six days; at the end of this time 87% preferred the moist air. They were then placed in flour that had been oven-dried to a moisture-content of 0·13% and kept for

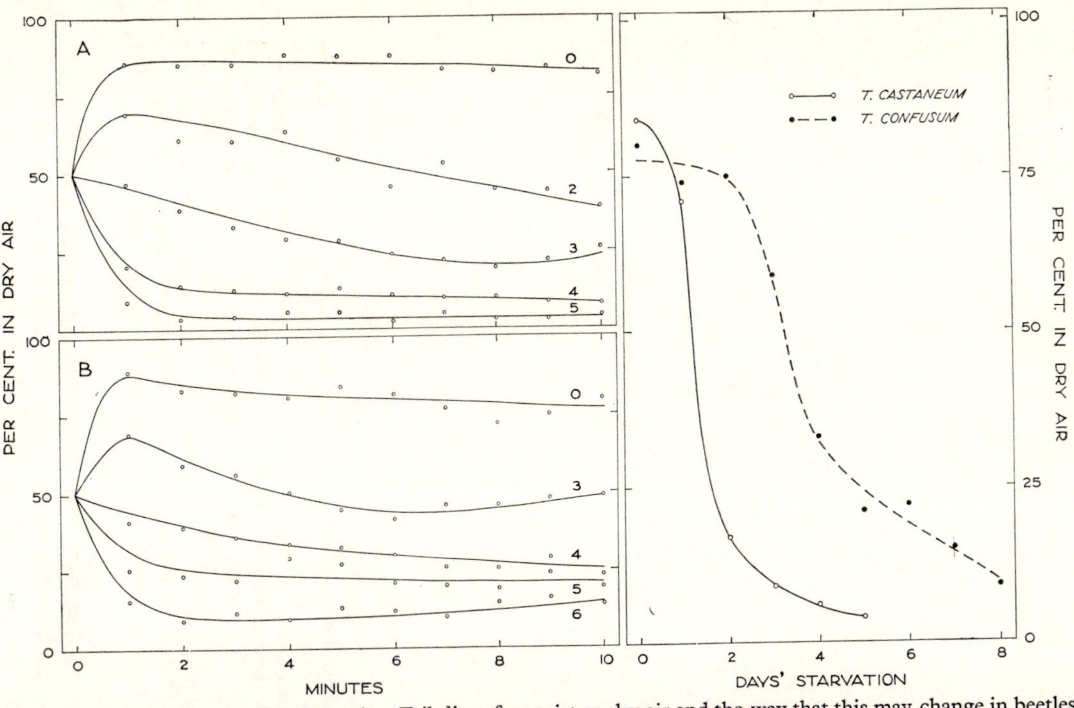

Fig. 6.11. The preference of adult beetles, *Tribolium*, for moist or dry air and the way that this may change in beetles that have been starved and desiccated. A, *T. castaneum*. B, *T. confusum*. The ordinates show the per cent of beetles on the dry side of the olfactometer. The figures against the curves show the number of days' starvation before the test began. The curves on the right show the period of starvation required to reverse the beetles' normal preference for dry air. (After Andrewartha and Birch, 1954.)

66 hours in this flour in a desiccator containing anhydrous calcium chloride. After this treatment 67% of the beetles preferred the dry air. The opportunity to eat, even though the food was oven-dry flour, had so altered the condition of the beetles that 54% of them had regained their normal preference for dry air (see *Ephestia*, sec. 6.2).

The moisture in the environment of an animal that is living in soil is best measured in terms of the force required to 'pull' water out of the soil. One way to measure this force is to place some damp soil on a porous plate from which is suspended a long tube, open at the bottom, containing water. The water in the tube must make good contact with the water in the soil so that the water in the tube is held against gravity by the surface-forces holding the water in the soil. Soil physicists use the logarithm of the height of this column of water as a measure of the moisture-content of the soil; it is sometimes called the pF of the soil but there is a modern trend to drop this name because the analogy with the pH scale has not proved very helpful (Richards, 1949). It is more useful to think of the height of this column of water as representing the 'negative pressure' of the water in the soil. Maelzer's (1958) experiments with the scarabeid *Aphodius howitti* illustrate the usefulness of this way of expressing the water content of soil.

The life-cycle from egg to adult of *Aphodius* occupies one year. The whole life-cycle is spent in the soil except that the adults come to the surface to disperse before burrowing into the soil again to lay their eggs. The pupae transform to adults during December which is mid-summer and approaching the height of the dry season in South Australia. They lie at the foot of burrows, about nine inches below the surface of the soil, until they are stimulated by an increase in moisture to emerge and disperse. Usually a thunderstorm in January or February may bring enough rain to stimulate the beetles. In seeking places to burrow into the soil again and lay their eggs the beetles congregate quite strongly in particular areas. Maelzer postulated that the strong preference shown by the beetles for particular areas was associated with the moisture-content of the soil and he did the following experiment.

A number of small boxes, open at the top, were filled with two sorts of soil, a sand and a clay-loam. The water-content of the sand varied from 2½ to 12% and that of the clay-loam from 17 to 37%. The boxes were arranged at random in a large cage and a number of female *Aphodius* were placed in the cage. Subsequently the numbers of beetles that had burrowed into each box were recorded. Table 6.04 shows that for each soil there was an optimum water-content which was not related to the amount of water in the soil but to the force with which the water was held in the soil: in the sand 79% of the beetles were in the range pF 2·9–3·5 and in the clay 66% were in the range 2·9–3·2; on this scale the 'permanent wilting coefficient' for plants is about 4·2.

Fraenkel and Gunn (1940) speak of a 'token stimulus' when the behaviour of an animal in response to a particular stimulus does not have survival value in relation to this stimulus but does have survival value in relation to some other component of environment which is indirectly related to the 'token stimulus'. In this sense moisture is a token stimulus to *Aphodius*.

Table 6.04.* *The number of female Aphodius (as per cent of total) found burrowing in soils of different water contents (expressed as per cent dry weight of soil and as the logarithm of the 'negative pressure' of the water in the soil)*

Type of soil	Moisture-content		No. of beetles (% total)
	%	Log. neg. press.	
Sand	11·5	2·3	15
	8·5	2·9	45
	3·5	3·5	34
	2·5	4·0	6
Clay-loam	37·0	2·5	22
	31·0	2·9	38
	25·0	3·2	28
	17·0	3·7	12

* After Maelzer (1958).

The explanation is as follows. The beetles usually oviposit during a temporary lull in the dry season brought about by a summer thunderstorm. If, as often happens, no more than 25–50 points of rain fall on soil that is 'air-dry' the places most likely to remain moist enough to attract the ovipositing beetles are relatively bare areas carrying the sparse stubble of winter-growing pastures rather than areas where the pastures are dense and include perennials, because a dense dry stubble may arrest 15–25 points of rain without it ever reaching the soil. In the former sort of place the pasture which develops during next winter and spring is not likely to make a dense sward; the feeding of the larvae is likely to make it sparser. Consequently, by the beginning of summer the soil where the mature larvae and subsequently the pupae and adults must lie dormant, waiting for the time when the adults may disperse, is likely to be moister than that under a dense sward of pasture, especially if it includes perennials which continue to draw on the water in the soil late into the dry season. Thus, although these insects are not very hardy they are fairly secure because the behaviour of the adult (response to a 'token stimulus') ensures that the larvae will be living in places that are relatively moist at the time when the risks from dryness are greatest.

The trap-door spider, *Blakistonea aurea*, is abundant in grassland and savannah on the plains near Adelaide. During the winter the spider feeds on whatever small arthropods it can catch near the mouth of its burrow. On the approach of summer it seals the lid of its burrow with silk and retires to the bottom of the burrow where it remains inactive without food or drink throughout the summer. At first sight it seems odd that the spider should shut itself off from supplies of food during the hottest and driest part of the year when the need to replenish water lost by evaporation is greatest. It seems especially odd in the light of experiments which showed that the spider is not much good at resisting evaporation from its body: when several spiders were exposed to dry air (relative humidity close to zero) they died in three days, having lost 30% of their original weight.

During the summer in Adelaide on hot days the maximum shade temperature may exceed 40°C and the relative humidity may fall below 10%. Hot dry weather is the rule for the five months from November to March. Yet most of the spiders that enter the summer are still alive at the end of it. The explanation of this paradox is that 20 cm below the surface the weather is very different from that near the surface; at the bottom of the burrow the relative humidity usually remains above 90%. Apparently it is safer for the spider to remain without food in a place where the air is moist than to seek food at the mouth of the burrow where the air is likely to be dry. The magnitude of this risk is doubtless a reflection of the spider's chancy way of seeking food – sitting at the top of the burrow waiting for an ant or beetle to blunder by.

The kangaroo rat, *Dipodomys merriami*, which lives in the desert in Arizona, also has a pattern of behaviour that protects it from exposure to the most extreme hazards of dryness in the desert. B. Schmidt-Nielsen and K. Schmidt-Nielsen (1951) showed that *Dipodomys* could maintain the water-content of its tissues in air at 24°C and 10–20% relative humidity with nothing to drink and no food other than air-dry barley.

During early summer when the weather had not yet reached its most extreme the temperature and humidity near the ground during the day varied between 20–45°C and 5–15% relative humidity. During the night the readings ranged between 15–25°C and 15–40% relative humidity. The kangaroo rat could not live for long if it were exposed to these daytime temperatures and humidities; but it spends the day curled up in a burrow that is about one foot beneath the surface of the ground, and emerges only at night to forage. B. Schmidt-Nielsen and K. Schmidt-Nielsen (1950) measured the temperature and humidity in the burrow during the day while it was occupied by the rat. They tied a very small thermohygrograph to the tail of a rat and let it go at the mouth of its burrow. Twelve hours later they excavated the burrow and retrieved the instrument. The maximum temperature varied from 25° to 30°C and the relative humidity from 30 to 50%. The absolute humidity varied from 8 to 15 mg per l. inside the burrow and from one to five mg per l. outside the burrow.

B. Schmidt-Nielsen and K. Schmidt-Nielsen (1949) calculated the amount of water that *Dipodomys* saved by staying in its burrow during the day:

> We can assume that *Dipodomys*, like other mammals, expires air that is saturated with moisture slightly below body temperature. Normal body temperature for *Dipodomys merriami* is 36–37°C and the following calculations are made under the assumption that the expired air has a moisture content corresponding to saturation at 33°C (Dill, 1938, p. 26). Air saturated with water vapour at 33°C contains 35·3 mg of water per l. of air. A *Dipodomys* that is breathing completely dry air will, therefore, lose 35 mg of water per l. of air expired. If the animal breathes desert air with a moisture-content, as we measured it during the daytime, of about 2 mg of water per litre of air, it will lose approximately 33 mg of water per l. of air expired. If, however, the animal stays in its burrow, the air it inspires will contain about 10 mg water per l., and the animal will lose only about 25 mg water per l. of expired air, or 24% less than in the first case.

6.22 *Physiological mechanisms for conserving water*

6.221 'Water-balance' in aquatic animals

Krogh (1939) discussed this subject thoroughly. I have taken the following brief summary largely from his book. It is also discussed by Ramsay (1952, Chap. 4).

Most marine invertebrates have the fluids in their bodies isotonic with the water outside. Some species that live in estuaries and in other places where the salinity of the water may vary widely also do this. They can live in such places by virtue of adaptations that allow them to carry on their life-processes when their cells are bathed in fluids with a wide range of osmotic pressures. Although the osmotic pressure of the blood may vary widely the concentration of particular ions, e.g. potassium, may remain more constant. So long as the concentration of salts in their blood remains within the range that they can tolerate the proportion of water in their bodies remains fairly constant. At extreme concentrations they die, but it seems more appropriate to think of them dying from imbalance of salts rather than from excess or lack of water.

All marine vertebrates and some invertebrates maintain their blood hypotonic to the water by excreting salts. The invertebrates in this group form a graded series from those in which the osmotic pressure of the blood is only slightly different from the water and usually responds to changes in the salinity of the water, to those in which the blood is quite markedly hypotonic to the water and may be largely independent of changes in the salinity of the water. In teleost fish salts are excreted chiefly by the gills. Water is lost with the urine and by osmosis mostly through the gills. It is replaced by drinking and with the food.

Nearly all the animals that live in fresh water absorb more or less water by osmosis and also gain some with their food. The whole of this has to be got rid of by excretion. The excretion of water tends to drain salts out of the body. Salts that are lost in this way may be replaced with the food. Certain Crustacea, e.g. *Daphnia*, can live for days in distilled water without food. This implies a remarkable capacity for extracting salts from the urine before it is excreted. Certain Crustacea, including *Daphnia*, can also absorb salts from extremely dilute concentrations in the surrounding water. In the mosquito larva the anal papillae are permeable to water and they also absorb salts from the water; the rest of the cuticle is highly impermeable (Wigglesworth, 1933; Koch, 1938). Krogh (1937) showed that a number of fresh water fish could absorb salts from the water. It seems that most of the animals that live in fresh water follow the same pattern. Many develop a cuticle that restricts the movement of water into the body by osmosis; and they maintain the water-content of their bodies and the osmotic pressure of the fluids that bathe their tissues fairly constant, by excreting the water that is gained by osmosis and with the food, by retaining the salts taken in with the food, and by absorbing salts from the water in which they are living. The adult insects that live in fresh water, e.g. *Hydrophilus*, are exceptions to this rule. They breathe air through the tracheal system and have the entire body covered by an impermeable cuticle that virtually prevents the entry of any water by osmosis.

The eel, the salmon and other teleost fish are remarkable because they move from the sea to fresh water and back again as a regular part of their life-cycle. While living in fresh water the eel drinks no water and secretes a copious dilute urine which contains scarcely any salts. In sea-water the eel does not eat but it replaces the water that is lost by osmosis by drinking sea-water; and it maintains a relatively constant osmotic pressure internally by excreting salts like other marine fish.

6.222 Conservation of water in terrestrial insects, ticks and snails

Insects that live in dry places may retain their nitrogenous waste-products as insoluble urates or they may excrete them with the faeces, which are dried out by special rectal glands (Wigglesworth, 1932). So virtually the only water that is lost by insects that live in dry places is that which evaporates through the cuticle or is lost from the spiracles during respiration.

Insects are remarkable because they lose so little water through the cuticle despite their small size. For example, the egg of the grasshopper *Austroicetes* is about 2 mm long by 1 mm dia. It weighs about 4·9 mg, of which 3·3 mg is water. While it remains in diapause it can survive the loss of 2·5 mg but will die if it loses much more than this. In nature, the eggs are laid in hard bare soil about half an inch below the surface; they remain dormant in the soil during the long, hot dry summer, hatching in the following spring. I have estimated that a standard Australian evaporimeter (which is a tank three feet in dia. sunk into the ground until its surface is level with the ground) would evaporate three feet of water while 50% of the eggs were losing enough water to kill them. That is, the average egg would have lost 2·5 mg while the evaporimeter was evaporating three feet. The morphology of the egg of *A. cruciata* is like that of the egg of *Locustana pardalina* which is also drought-hardy.

Matthee (1951) found that the extreme drought-hardiness of the egg of the locust *L. paradalina* is partly explained by the relative impermeability of the cuticle that covers the entire egg except for a small permeable 'hydropyle' at the posterior pole, through which water is taken up when the soil is moistened by rain. The impermeability of the cuticle is largely due to a layer of wax incorporated deep in the cuticle. The impermeability of the cuticle of insects and ticks is also due to a layer of wax which usually forms the outermost layer but one in the epicuticle; it may be covered and protected by a thin layer of 'cement' (Wigglesworth, 1948).

Lees (1947) measured the rate at which water evaporated through the cuticle of several species of ticks. The ticks were dead and the openings of the respiratory and alimentary systems had been blocked with wax; so it is likely that most of the water was lost through the cuticle. Figure 6.12 shows that at moderate temperatures the cuticles of all the ticks transmitted rather less than 1 mg of water per cm^2 per hr. But for each species there was a narrow range of temperature in which the rate of evaporation increased enormously. In general this critical zone

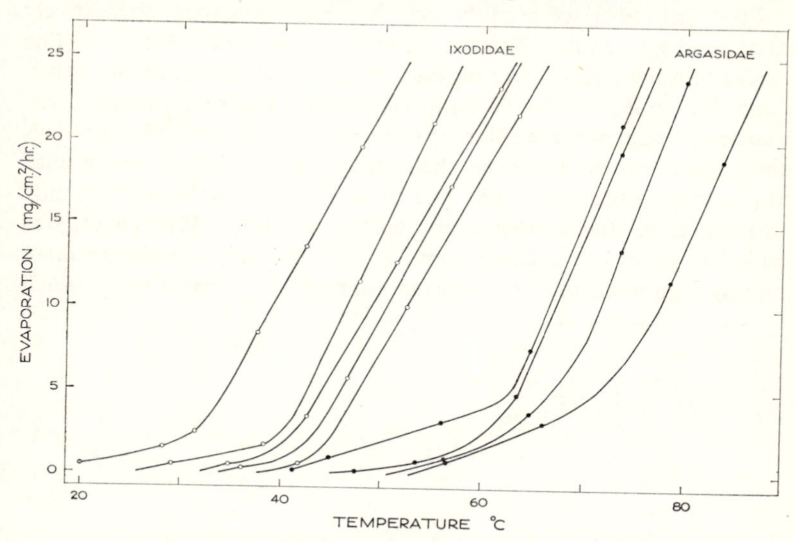

Fig. 6.12. The rate of evaporation of water from dead engorged ticks at different temperatures. The curves refer, from left to right, to the following species: *Ixodes ricinus, I. hexagonus, Amblyomma americanum, Dermacentor andersoni, Hyalomma savignyi* (Ixodidae), *Ornithodorus moubata, Argas persicus, Ornithodorus acinus, O. savignyi* (Argasidae). (After Lees, 1947.)

occurred at higher temperatures in the Argasidae which normally live in much drier places than the Ixodidae. Lees showed that these differences were associated with waxes of different melting points. (Compare Fig. 6.09.)

But the water-balance of an insect or a tick that is starving in dry air is not to be explained merely in terms of the physical permeability of the cuticle. Insects and ticks can use the energy of respiration to prevent evaporation from their bodies or to absorb water-vapour from air against substantial gradients of pressure. Lees (1946) exposed the tick *Ixodes* to air at 25°C and 50% relative humidity until they had lost about 10–20% of their original weight. He then placed them at 95% relative humidity and found that they rapidly regained their original weight by absorbing water-vapour from the air. By a series of similar experiments he found that the critical humidity for *Ixodes* was about 92%. The ticks could be starved in air at 92% relative humidity without losing water; alternatively ticks that had been desiccated could absorb water-vapour from the air in which the relative humidity exceeded 92%.

Browning (1954) exposed the tick *Ornithodorus moubata* alternatively to 0% and 95% relative humidity at 25°C as indicated in Fig. 6.13. The ticks lost weight at 0% and regained it rapidly at 95% humidity; they continued to do this for 60 days without any apparent change in the maximal weight achieved during exposure to moist air. After 60 days the pattern was the same but the maximal weight fell off slowly until after 140 days they died. The first cause of death was almost certainly exhaustion of 'fuel' rather than shortage of water. When CO_2 was added to air at 0% relative humidity, the ticks lost weight no more rapidly than in air until the concentration of CO_2 reached 30%, which

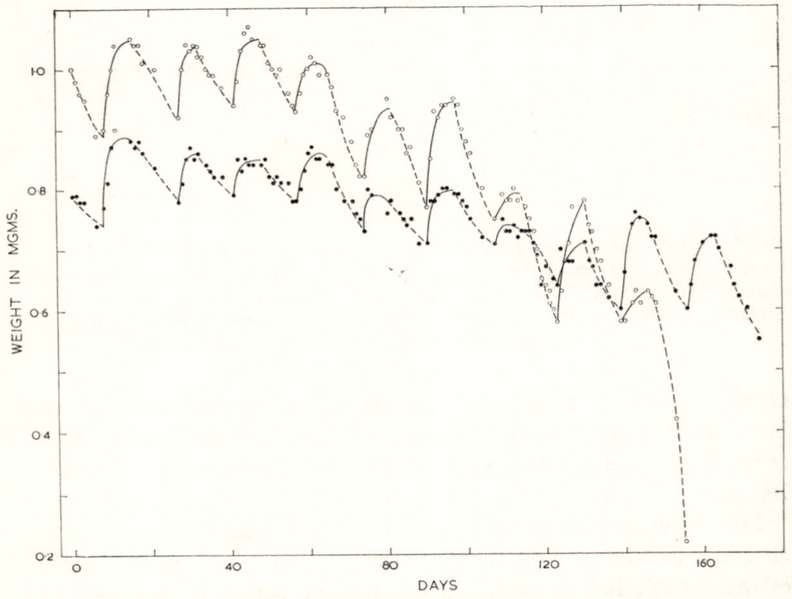

Fig. 6.13. Changes in weight of two second-instar nymphs of *Ornithodorus moubata* that were exposed alternately to 95% relative humidity (complete lines) and 5% relative humidity (broken lines). (After Browning, 1954.)

is the concentration that is required for anaesthesia. Similarly in mixtures of air and nitrogen the ticks retained their normal ability to resist desiccation until the concentration of nitrogen approached 98%, suggesting that the failure of the ticks was caused by shortage of oxygen.

Fraenkel and Blewett (1944) reared larvae of the moth *Ephestia* in three lots of flour with water-contents 1·1%, 6·6% and 14·4%. The three lots of flour were in equilibrium with air having relative humidities 0%, 20% and 70%. Table 6.05 shows that the pupae reared in dry flour contained nearly the same proportion of water as those reared in moist flour, but those in the dry flour digested more food per mg of tissue produced. At least some of the water in the pupae that had developed in the moist flour may have come from the food, but virtually all the water in the pupae in the dry flour must have been manufactured in the body by oxidizing food. This means that the larvae were able to respire carbon dioxide while retaining some of the water of respiration.

Table 6.05*. *Relationship between moisture-content of food and amounts of food digested to produce the same weight of dry matter in the tissues of Ephestia kuhniella reared in flour at 25°C*

Moisture in food (%)	Relative humidity (%)	Wet weight of pupa (mg)	Water-content of pupa (%)	Ratio dry wt. food to dry wt. pupa
14·4	70	25·5	68	6·3
6·6	20	18·7	66	9·0
1·1	0	15·8	64	12·7

* After Fraenkel and Blewett (1944).

I do not know what proportion was retained because Fraenkel and Blewett did not record the humidity of the air that was expired. Lees (1947) showed that the tick *I. ricinus* could not absorb water-vapour from moist air while the cells of the epidermis, lying immediately below the cuticle, were occupied repairing even a small area of cuticle which had been abraded with fine grit to remove the wax. From the results of these and other experiments Lees concluded that the site of water-absorption in the tick was in the cellular epidermis.

On the other hand, Walters (1966) working with the larvae of the mealworm, *Tenebrio molitor*, has shown that the power (i.e. rate of doing work) which the grubs devote to water-conservation is independent of humidity over a wide range of relative humidities; but it is closely dependent on temperature. The results of these and other experiments suggest that in *Tenebrio* water-conservation may be regarded merely as a by-product of respiration. The corollary is that the grubs of *Tenebrio* expire air that has been dried to about 90% R.H.; but this hypothesis has not been confirmed empirically. If this is true for *Tenebrio* then it is also likely that the camel tick expires air that has been dried to 85% R.H. and the prepupa of the flea expires air that has been dried to 50% R.H. (see Table 6.08). If this hypothesis can be verified it would suggest a mechanism for water-conservation quite different from any that is known in vertebrates. The camel and the kangaroo rat both expire water that is saturated with water-vapour at the temperature of their lungs – or at least at that of their nasal passages.

Unlike the ticks and the insects, the snail *H. virgata* seems not to be able to absorb water-vapour from humid air – its flair seems to lie in holding tenaciously to what it has already got (Pomeroy, 1966). That this is an active process which goes on only while the snail remains alive is strongly suggested by Fig. 6.14 which describes the changing weight of a snail that was kept continuously at 30°C in air with a relative humidity of 5%. The sharp inflexion in the curve which occurred about the 190th day coincided with the death of the snail.

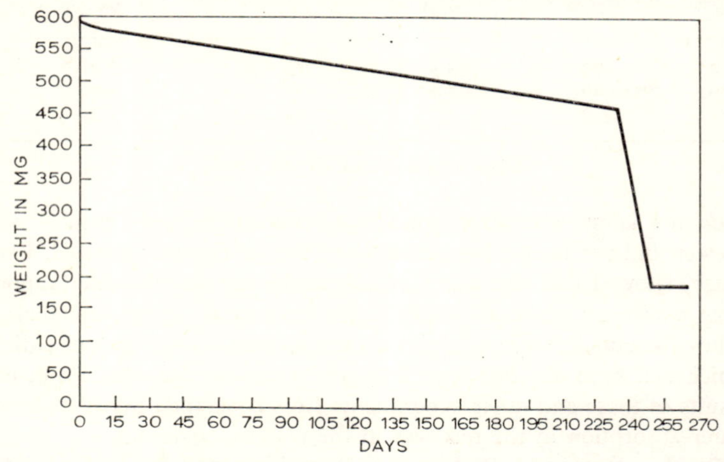

Fig. 6.14. The snail, *Helicella virgata*, suspended in air of 5% relative humidity at 30°C, lost weight (water) slowly until it died after 234 days.

Similar curves could be drawn for the other 39 snails in the experiment. The average for the whole experiment is shown in Table 6.06.

At the beginning of the experiment a snail contained on the average 407 mg of water; by the time of death this had fallen to 271 mg.

The figures in Table 6.06 allow us to speculate on the amount of work that a snail may have done (during the 234 days that it was living at 30°C in air of relative humidity 5%) in order to conserve 271 mg of the 407 mg of water that it had in its body at the beginning of the experiment. If we assume that the abrupt change in the daily rate of water-loss that occurred about the 234th day was caused by the death of the snail (Fig. 6.14) we can reasonably argue that the living snail was doing some sort of work that resulted in it not losing 19·830 – 0·559 = 19·271 mg of water a day. Now consider the analogy of a man who is pumping water into an overhead tank with a small hole in it from which

Table 6.06. *The mean daily loss of weight (water) by 40 snails kept at 30°C in air of 5% relative humidity*

Mean weight at beginning of aestivation	Mean daily loss of weight during life	Mean daily loss of weight after death	Mean duration of life
mg	mg	mg	days
592	0·559	19·830	234

the water is running away. If the hole is dripping at the rate of 19·830 gal/hr and the amount of water in the tank, despite the man's pumping, is decreasing at the rate of 0·559 gal/hr, by analogy the snail may be considered to be doing the equivalent of pumping 19·271 mg of water per day into its body.

The work done by a pump is given by the equation:

$$E = Pv$$

where E is energy, P is pressure, and v is volume of fluid pumped against the pressure P. The pressure can be calculated from the equation given above; the volume, taking the density of water as 1, is 0·0198 cm³ per day or 4·5 cm³ in 234 days. The amount of energy required to 'pump' water from a place where the relative humidity is 5% (equivalent to an osmotic pressure of about 4,000 atmospheres) to a place where the pressure is eight atmospheres (i.e. the body-fluids of the snail) is about 0·5 kilo-calories which would require the oxidation of at least 20 times as much dry matter as the snail actually uses during its period of aestivation. Whatever the mechanism the snail may have for conserving water during aestivation, clearly there is no point in comparing it with a pump.

Yet it would seem to be some sort of active process. In addition to the evidence of the shape of the curve in Fig. 6.14, Pomeroy's experiments have provided the following facts:

(*a*) The conservation of water breaks down in the absence of oxygen.
(*b*) When experiments, like the one from which Fig. 6.14 was constructed, were repeated at various combinations of temperature and humidity it turned out that the duration of life was nearly independent of humidity but closely related to temperature.
(*c*) The rate of loss of water depended on humidity, but even at the lowest humidities the snail seemed to have more than ample power (rate of doing work) to retain a safe amount of water in its body.
(*d*) When it eventually died it seemed that death came from shortage of food rather than shortage of water.

All these results, taken together, suggest an active (i.e. energetic) mechanism for water-conservation. It does not seem as if the energy is used in a process that has any resemblance to pumping; perhaps the work that the snail does is more like the process of mending holes in a ready-made wall.

6.223 Conservation of water in terrestrial mammals

Small rodents such as the American kangaroo rat, *Dipodomys merriami*, or the Australian hopping mouse, *Notomys alexis*, are the most drought-hardy mammals known. In nature they live in deserts eating air-dry seeds and drinking no water. In the laboratory *N. alexis* gained weight and water living at 25°C in air of relative humidity 35% and eating dry birdseed with a moisture content of 10% (MacMillen and Lee, 1969). The kangaroo rat also gained weight on a diet of dry barley and no water; when it was offered no other food than one that was high in nitrogen, as soy bean, it drank; when no other water was available it drank sea-water; it lived for weeks, and even gained weight, on a diet of soy bean and sea-water.

The physiology of *Dipodomys*, especially with respect to the conservation of water, has been studied by K. Schmidt-Nielsen and B. Schmidt-Nielsen (1952 and other papers). They first did experiments which demonstrated that (*a*) *Dipodomys* did not store excretory wastes as harmless insoluble solids in its tissues. (*b*) It did not carry in its body a reserve of water that could be drawn on in an emergency. (*c*) It did not survive an unusual reduction in the water-content of its body (section 6.2). In short, *Dipodomys* used water in just the same way as any other mammal (except that it did not sweat) but it was very economical.

Table 6.07 compares the water-economy of *Dipodomys* with that of *Rattus* and man. The difference between *Dipodomys* and *Rattus* in the amount of water lost by evaporation seems too big to be accounted for by the little sweating done by *Rattus*. There was no evidence from breathing of *Dipodomys* or from the oxygen-dissociation curve of its blood to suggest that it used oxygen any more efficiently than *Rattus*.

Table 6.07. *Water-economy of Dipodomys compared with that of Rattus and man*

	Dipodomys	*Rattus*	Man
Evaporation from all sources (mg. H_2O per ml. O_2)	0·54*	0·94†	‡
Concentration of urine (urea)	3·8M§	2·5M	1·0M
Total electrolytes	1·5N‖	0·6N	0·37N
Water-content of faeces	45%	68%	
Water lost with faeces per 100 g barley	2·5 g	13·5 g	

* None lost by sweat.
† A little lost by sweat.
‡ May lose 10 times as much from sweat as from all other sources combined.
§ Equivalent to 23% urea.
‖ Equivalent to 8·5% NaCl.

But it was suggested that the shape of the nose in *Dipodomys* may be an adaptation for conserving water. The nose is long and thin and poorly supplied with blood. When the temperature of the air is below 37°C (the body temperature of *Dipodomys*) the temperature of the nasal passages may be lower than the temperature of the lungs which may allow some of the water of respiration to condense and be reabsorbed in the nose.

The total intake of water comprises that extracted from the food, that manufactured by oxidizing the food, and that inhaled during breathing. The total output of water comprises that evaporated from skin and lungs, that excreted with the urine and that excreted with the faeces. Because all these quantities were known, or could be calculated, K. Schmidt-Nielsen and B. Schmidt-Nielsen (1952) were able to draw up a complete 'balance sheet' of the water-economy of *Dipodomys*, which I have reproduced in Fig. 6.15. The lines for intake and output intersect at 10% relative humidity which is the lowest humidity at which *Dipodomys* can live at 25°C when it has only dry barley for food and no water to drink.

Elsewhere in deserts widely separated from each other there are many species of small rodents that look and behave like *Dipodomys*.

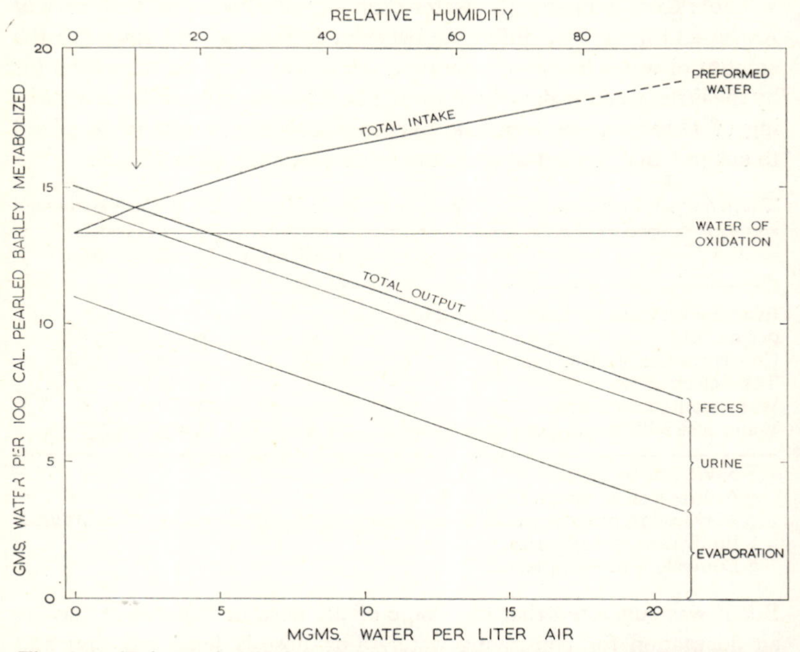

Fig. 6.15. 'Balance sheet' of the water-economy of *Dipodomys*. For explanation see text. (After Schmidt-Nielsen and Schmidt-Nielsen, 1951.)

They have long hind legs and long narrow snouts; they proceed by jumping; they live underground in burrows and are nocturnal; and their staple diet is air-dry seeds. In America they are heteromyids; in Africa and Europe they are cricetids and dipodids; in Australia they are murids. The convergence is striking confirmation for Ewer's (1968) argument that anatomy, physiology and behaviour evolve together. The explanation lies in ecology.

The camel is well known for its ability to live and work in the desert (Schmidt-Nielsen *et al.*, 1956). The camel is too large to seek the shelter of an underground burrow. It often has to live in places that are much hotter than itself, and where it is exposed to direct radiation from the sun. Even when it is in a place that is slightly cooler than itself the camel, because of its small surface area relative to its weight (a consequence of large size), cannot dissipate enough heat merely by radiation: it must supplement radiation by the evaporation of sweat. Thanks to a series of elegant studies by Schmidt-Nielsen and his colleagues, we now know how the camel can do so well in the desert despite these seeming difficulties. The following summary is taken from Schmidt-Nielsen (1964).

Compared to a man or a dog, a camel can lose proportionally more water from its body without dying. If a man loses more water than about 10% of his original weight he becomes incapable of looking after himself. As he loses still more water his blood becomes viscous and no longer flows freely enough to transport heat from deep tissues to the surface where it can be dissipated. As the loss of water approaches 18% the temperature probably rises explosively and the man probably dies quickly. By contrast a camel can lose water at least equal to 20% of its weight and still have its blood non-viscous and circulating freely. This is because the water that the camel loses comes largely from its tissues, whereas the water that a man loses comes largely from his blood.

On a warm day a man will start to sweat as soon as his temperature begins to exceed 37°C; nor will his temperature fall much below 37°C even on a cold night. By contrast the temperature of a camel, especially one that is dehydrated, will fall, during the night, as low as 34°C, and rise during the day to about 40°C. Sweating does not begin until the body temperature exceeds 40°C. Schmidt-Nielsen has estimated that an average-size camel would require about 2,500 kilo-calories to raise its temperature through 6°C; to dissipate this much heat by sweating would require the evaporation of five l. of water. By a carefully controlled departure from strict homoiothermy the camel saves five l. of sweat a day.

The curly fur of a camel traps air which insulates the animal against the heat radiating from the sun and the hot ground. Also the fur ensures that when the camel is sweating the water will evaporate close to the skin where it can do most good.

The adaptations for water-conservation in the camel are made possible by its large size just as the adaptations in the kangaroo rat are related to its small size. The invertebrate animals that I have included in Table 6.08 are much smaller than the kangaroo rat and they are poikilothermic. Because they are poikilothermic and because in such small animals metabolic heat is readily dissipated by radiation they do not need to expend any water for evaporative cooling. Ticks and insects have cuticles that incorporate wax which makes them highly impermeable to water; they use almost no water for excretion because nitrogen is excreted as uric acid and the faeces are dried out in the rectum before being excreted; in addition some of the arthropods can absorb water vapour from moist air.

There is no useful common denominator by which these adaptations might be compared quantitatively, but they do emphasize the importance of anatomy. For example, the gull (Table 6.08) excretes salt from a specialized nasal gland (if you watch a gull for a few minutes you will probably see it shaking drops of concentrated salt solution from its nose); but it also uses its kidneys in the normal way for excreting nitrogenous waste. Because in the camel and the kangaroo rat the kidneys have to excrete both salt and urea, their kidneys have to work against greater gradients in pressure to achieve the same result – it is a matter of design, like the difference beween two-wheel and four-wheel drive on a car. The literature does not contain enough information to allow comparisons based on work or power (rate of doing work). But the different sorts of animals can be compared quantitatively by reference to the pressure against which the animal retains water on one side of a membrane or moves water across membranes in its body. The equation which allows measurements of the activity of water taken in one scale to be transformed to another is:

$$\pi = -\frac{RT}{v}\log_e(1 - X)$$

$$= \frac{RT}{v}\log_e\frac{P_0}{P}$$

Where π = osmotic pressure in atmospheres
R = 0·08206
T = absolute temperature
v = volume of 1 mole of solvent (for water v = 0·018)
X = the mole fraction of the solute
P_0 = vapour pressure of solvent
P = vapour pressure of solution
$\frac{100P}{P_0}$ = relative humidity of air in equilibrium with the solution

This equation has been used to calculate the figures in Table 6.08.

Table 6.08. *The 'activity' of water against which certain animals work to conserve water in their bodies. 'Activity' is measured in atmospheres of osmotic pressure*

Species	Empirical information	'Activity' atmospheres pressure	Author
Rat flea	Gains weight in air R.H. 50%	910*	Edney (1947)
Brine shrimp	Maintains body-fluid in saturated brine	326†	Croghan (1958)
Camel tick	Gains weight in air R.H. 85%	216*	Lees (1947)
Hopping mouse	Concentrates urine to 6·6 osmoles/l	147‡	MacMillen and Lee (1969)
Kangaroo rat	Concentrates urine to 5·5 osmoles/l	123†	Schmidt-Nielsen (1964)
Camel	Concentrates urine to 2·8 osmoles/l	63†	,, ,, ,,
Man	Concentrates urine to 1·4 osmoles/l	32†	,, ,, ,,
Gull	Excretes concentrated solution of NaCl	40†	Schmidt-Nielsen and Sladen (1958)
Eel	Excretes salts into sea water	26*	Krogh (1939)
Mosquito larva	Extracts salts from fresh water	8	Wigglesworth (1933)

* Converted from measurements of relative humidity.
† Converted from measurements of depression of freezing point.
‡ Estimated with Kolber biological cryostat.

6.3 LIGHT

The exciting aspect of light, to the animal ecologist, is the diverse ways in which animals have become adapted to respond to light as a 'token stimulus', especially the ways in which it serves as a 'clock' by which life-cycles are synchronized with each other or with the seasons. Physiological studies, especially of the endocrine organs of mammals, birds and insects, have greatly enhanced our understanding of light as a component of environment just as our understanding of moisture has been helped by physiological studies of osmoregulation and active transport.

6.31 *The influence of light in synchronizing life-cycles with each other*

The cockroach *Periplaneta americana* is nocturnal. If it is living in a place where daylight penetrates it will remain quiet during the day and become active at night. But when a number of *Periplaneta*, that had previously been exposed to the natural diurnal rhythm of light, were kept in continuous darkness they continued to be quiet during the day and active at night for about four days. After this their behaviour lost its regular rhythm (Harker, 1956). When another batch was exposed to artificial light during the natural night and kept in darkness during the natural day they were quiet during the night and active during the day; and this rhythm also persisted for several days when they were kept in continuous darkness. By grafting a legless cockroach that had been conditioned for a certain rhythm on to the back of another that had lost its rhythm (by living in continuous light) Harker showed that the rhythm was caused by a hormone circulating in the blood.

Bateman (1955) showed that a rhythm imposed on one generation of the fruit-fly *Dacus* was transmitted to the pupae of the next generation. He illuminated two batches of adult flies as follows: one batch was kept in the light from 9 a.m. to 5 p.m. and in the dark for the rest of the day; the other batch was illuminated from 9 p.m. to 5 a.m. The larvae and pupae were kept in continuous darkness, and the adults of the next generation emerged in darkness. Most of the emergences occurred during a period of 48 hours but the emergences were not distributed evenly over this period; on the contrary there were two peaks of emergence separated by 24 hours; the striking thing about them was that they occurred in each batch during that time of the day when the adults of the previous generation had been illuminated. This remarkable example of a 'remembered' rhythm is illustrated in Fig. 6.16. I expect

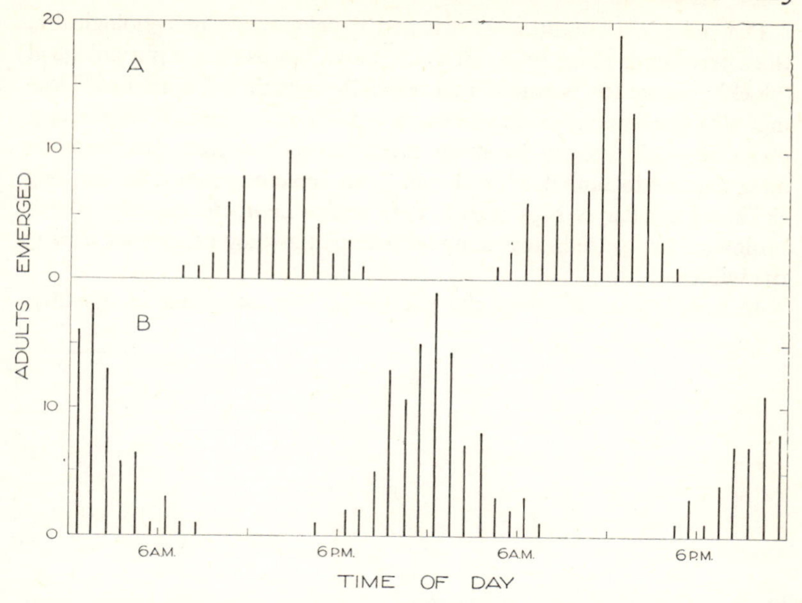

Fig. 6.16. The influence of exposure to light during the adult stage on the 'timing' of the pupal ecdysis in the next generation, in *Dacus tryoni*. A, adults were illuminated from 9.00 a.m. to 5.00 p.m.; B, adults were illuminated from 9.00 p.m. to 5.00 a.m. (After Bateman, 1955.)

that this rhythm will also be found to be controlled by hormones, but at present we know nothing of the mechanism.

6.32 *The influence of light in synchronizing the life-cycle with the season of the year*

The caterpillars of *Grapholitha molesta* are serious pests of peaches in certain parts of the world. One of the reasons why *Grapholitha* can maintain such high numbers, except when it is sprayed with insecticides or 'controlled' by predators, is that its life-cycle is nicely synchronized with the seasons and with the life-history of its food-plant the peach. During summer there may be several generations and all stages of the life-cycle may be present at once. But during winter when there is no food and when temperature may be low enough to prevent development or even to kill the active stages, the whole population comprises only fully mature caterpillars in a dormant cold-hardy condition which is known as 'diapause' (Dickson, 1949).

Diapause is a condition of arrested development: morphologically, the caterpillar is ready to moult to the pupa, but there is a physiological 'block' because the hormone that normally organizes the moult is lacking. The hormone is lacking because a particular group of neurosecretory cells at the base of the brain which normally secrete this hormone have become incompetent to do so. They remain incompetent so long as the caterpillar is kept warm; they regain competence only after a prolonged exposure to low temperature such as they experience during the winter.

In addition to this physiological 'block' the caterpillar in diapause

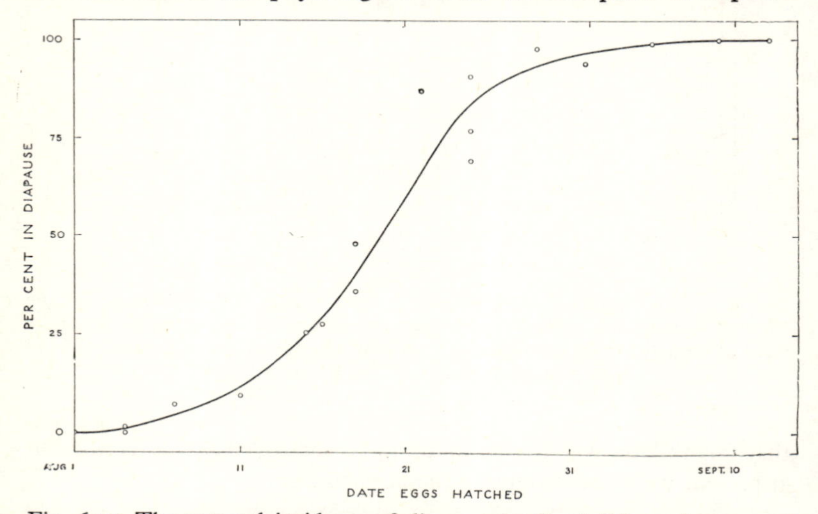

Fig. 6.17. The seasonal incidence of diapause in *Grapholitha molesta* reared out-of-doors at Riverside, California. (After Dickson, 1949.)

also shows certain other qualities – there is a smaller proportion of water in its body, it is more cold-hardy (sections 6.12, 6.2), and it develops a particular pattern of behaviour which leads it to seek a well-sheltered place in which to spend the winter. Diapause is found in many species of insects, mites and ticks, and an analogous condition is found in Crustacea, nematodes, rotifers, tardigrades, earthworms, snails and perhaps frogs.

The life-cycle of *Grapholitha* is so nicely synchronized to the seasons because the young caterpillars respond to length of day in a particular way: it seems as if the shortening days of autumn serve as a 'clock' to predict the approach of winter. Dickson (1949) collected young larvae from orchards in southern California, reared them and recorded the proportion entering diapause. Figure 6.17 shows that virtually all those that hatched from eggs after about August 25 entered diapause. At this time of the year there was little more than 12 hours of daylight each day. Dickson then did a series of experiments at 24°C varying the 'length of day' artificially from 0 to 24 hours. The results, shown in Fig. 6.18, confirm the observations that were made in the field.

Rowan (1938) kept the finch *Junco hyemalis* in outdoor aviaries exposed to the weather. One aviary was artificially lighted by two 50-W globes

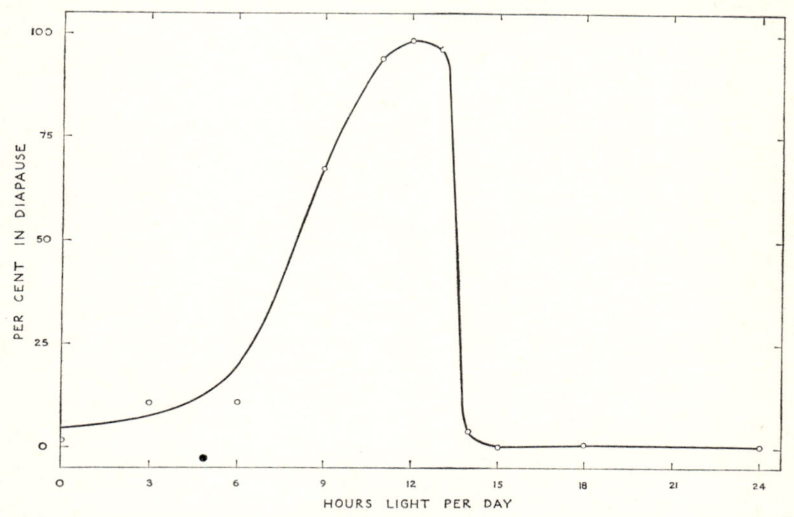

Fig. 6.18. The influence of photoperiod at 24°C on the inception of diapause in *Grapholitha molesta*. (After Dickson, 1949.)

which, from October, were switched on at sunset and each day were left on for five minutes longer than the day before. By mid-November the testes of the birds in this aviary were increasing in size and by the end of December they were larger than those of normal birds early in spring, despite the fact that the temperature in the aviary had been low – below 0°C at times. The testes of the birds that were not receiving additional light were still quite small. These results have since been confirmed by a number of similar experiments on different species (Marshall, 1942; Bullough, 1951).

Baker and Ransom (1932) found that the field-mouse *Microtus* bred freely when exposed to 15 hours of light each day but virtually ceased breeding at 9 hours. Bissonnette (1935) exposed ferrets to gradually increasing length of day and found that they became sexually mature more rapidly than the controls. Similar results have been reported with hedgehogs, raccoons, lizards, turtles and certain species of fish (Marshall, 1942; Bullough, 1951).

7 *Components of environment: malentities*

7.0 INTRODUCTION

Andrewartha and Birch (1954) recognized four components of environment one of which they called 'a place in which to live'. Andrewartha and Browning (1961), analysing the idea of resources in animal ecology, suggested that many of the interactions that had been discussed under a place in which to live might be properly encompassed by the idea of resources. And Browning (1962, 1963) suggested a fifth component of environment, *hazards*, which accounted for most of the interactions remaining in Andrewartha and Birch's discussion of a place in which to live after resources had been abstracted from it. This was a more elegant analysis because it made the components of environment more homogeneous and made the classification more functional: it separated 'resources' which have a positive influence on an animal's chance to survive and reproduce (Fig. 3.01) from 'hazards' which have a negative influence (Fig. 7.01.)

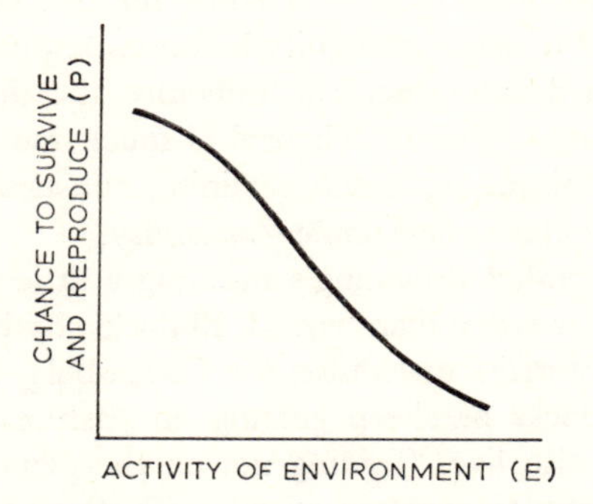

Fig. 7.01. An animal's chance to survive and reproduce is reduced when malentities are numerous. *E* stands for the activity of the environment (malentities); *P* stands for the animal's chance to survive and reproduce.

7.1 BROWNING'S DEFINITION OF HAZARDS

Browning described several hazards. For example, a sheep, carrying a number of ticks, *Ixodes ricinus*, might use a fencing post, a tree-trunk or an upstanding rock as a 'rubbing post' against which to squash or rub off the ticks. The presence of such rubbing posts reduces the ticks' chance to survive and reproduce. Or, bullocks, walking in heavy clay soil when it is wet, may leave deep hoofprints which may persist long after the bullocks have gone. In wet weather the hoofprints may fill with water. Springtails (Collembola) often become trapped and die in enormous numbers in the surface film in such little pools.

The hoofprints were classified as hazards because the springtails do not enter into the environment of the bullock whose chance to survive and reproduce is independent of the numbers of springtails trapped in the hoofprints. Browning emphasized this point by contrasting the hoofprints left by the bullock with the pit dug by an antlion (larva of Myrmeleontidae) or the web of a spider which are best classified, along with the gun of a hunter or the claws of a cat, as extensions (attributes) of the predator which enhance its chance of catching prey.

Browning conceived a hazard as a non-living material object which was inimical to the animal in whose environment it occurred. So it is quite clear that, in his usage, the hoofprints and not the bullocks are hazards.

7.2 MALENTITIES

Browning's choice of 'hazard' as a name for this component of environment has led to some ambiguity because, despite the dictionaries' definition of hazard as 'obstacle' or 'difficulty' the alternative usage in which hazard means 'risk' or 'chance' is much more popular in the writings of ecologists. So, while retaining substantially Browning's meaning, I have coined a new name – *malentity*.

I have also extended Browning's meaning a little so that malentity has a slightly wider scope than hazard. Bullocks, by leaving hoofprints behind them, may make malentities for Collembola, but elsewhere, in drier places, bullocks or sheep grazing on pastures where the snail *Helicella virgata* also lives (feeding not on the plants but by rasping organic matter from the surface of the soil) often, quite accidentally, tread on a snail and crush it; so also do men walking in parks which are much favoured by the snails. The snails do not enter into the environment of the animals that crush them whose chance to survive and reproduce is independent of the number of snails that are crushed. Browning excluded living animals from his concept of hazards because he had already considered them under another heading, 'members of other species', which he had adopted from the usage of Andrewartha and Birch (1954). But in the usage that I have adopted in this book it would seem artificial not to recognize the essential similarity (in function) of 'rubbing posts', 'hoofprints' and 'trampling animals'. So I include all three as malentities. I mention below several other examples of malentities which serve to broaden this analogy without exhausting its diversity.

Ullyet (1950) reared the larvae of three species of blowflies *Lucilia sericata*, *Chrysomyia chloropyga* and *C. albiceps*, either one species at a time or two together with different degrees of crowding on the same amount of meat. When two species were reared together Ullyett always placed equal numbers of each species on the meat at the beginning of the experiment. There is enough food in 140 g of meat to rear about 2,000 blowflies of normal weight. Ullyett varied the number of maggots on 140 g of meat between 100 and 10,000; at one extreme the crowding was negligible, at the other it was intense. When any one species was alone the consequence was as described in section 3.2; up to about 2,000 the death-rate was low and the survivors were of normal weight. As the crowding increased the weight of the individual survivors became less but the death-rate remained low; as crowding intensified there came a stage when no further diminution in weight was possible and the death-rate increased steeply. When *L. sericata* and *C. chloropyga* were together the decline in weight and increase in death-rate was parallel in the two species; they seemed to be competing on equal terms. But when *L. sericata* and *C. albiceps* were together the response of *L. sericata* was much stronger than that of *C. albiceps*; before the crowding had become very extreme the survival-rate in *L. sericata* had declined to zero. Some *L. sericata* were killed and eaten by *C. albiceps* but this seemed secondary to the general inability of *L. sericata* to thrive in the presence of many *C. albiceps*. Although the interaction between the two species was not analysed precisely it seems clear that *C. albiceps* was functioning as a malentity in the environment of *L. sericata*.

Waterhouse (1947) observed a similar interaction between *Chrysomyia rufifaces* and several other species of blowflies living in sheep carcases exposed out-of-doors near Canberra. During winter *Calliphora stygia* bred uncrowded in the carcases. During spring the carcases were crowded largely by larvae of *Lucilia sericata*, and *Calliphora augur*. During summer *Chrysomyia rufifaces* was abundant and the carcases were nearly always crowded with its larvae. During spring and summer *Lucilia cuprina* laid many eggs on the carcases but virtually none survived to maturity in a carcase that was crowded with any of the other species, especially *C. rufifaces*. A carcase that was crowded with *C. rufifaces* produced relatively few *L. sericata* or *C. stygia*. When a carcase was crowded with *C. rufifaces* the other species that were present especially *L. cuprina* seemed to be irritated and disturbed and many left the carcase and died before they were fully grown. It seems that, as in Ullyett's experiments, *C. rufifaces* was functioning as a malentity in the environments of the other species, especially *L. cuprina*, and that the other species were functioning as malentities in the environment of *L. cuprina*. One important consequence these malentities have in the environment of *L. cuprina* seeking to breed in carrion is that *L. cuprina* has nowhere to breed successfully except on the living sheep.

8 *Components of environment: more about the ecological web*

8.0 INTRODUCTION

In chapters 3–7 it was appropriate to mention a number of interactions which, although they did not involve the animal directly, nevertheless influenced its chance to survive and reproduce through their influence on the activity of a component of environment. Because their influence is indirect, these interactions were considered to be taking place in the outer reaches of the ecological web. In this chapter I develop this idea a little further.

The members of a sparse population may run a considerable risk of not meeting a mate (chapter 4). In section 8.1 I discuss certain other disadvantages of being a member of a sparse population; because these disadvantages involve interactions between the density of the population and various components of environment they are best discussed in relation to the ecological web. Intraspecific aggression consequent on crowding was discussed in section 5.3; the consequences of crowding relative to resources or predators were discussed in chapters 3 and 5. Because these interactions in the ecological web may be important in generating negative feed-back in natural populations they will be mentioned again in chapter 9 instead of in this chapter. It is quite usual for interactions generated in the second or more remote reaches of the ecological web to have an important influence on an animal's chance to survive and reproduce. In chapter 3 I mentioned how weather may modify the severity of an extrinsic relative shortage of food for *Thrips imaginis*; in section 8.2 I elaborate this point. Sometimes as the interactions in the ecological web extend into the outer reaches of the web they become very complex (section 8.3). Ecological barriers to the compact distribution of a resource may acerbate an extrinsic relative shortage, so such barriers are appropriately considered as part of the ecological web influencing the activity of a resource (section 8.4).

8.1 SOME ADVANTAGES IN BEING ONE OF MANY

The grain borer, *Rhizopertha dominica*, will thrive in a jar of wheat in the laboratory, but a new colony cannot be started with a few animals in wheat that is whole. If the wheat contains some grain that is cracked or broken few colonists will suffice; but if the grain is sound many are needed. When the insects are first placed with sound grain they begin to nibble it. When many insects are placed in the same jar this nibbling proceeds to the point where at least some grain has been sufficiently damaged to become favourable for *Rhizopertha* before all the insects have died; when only a few insects start with sound grain all of them are likely to die before enough grain has been adequately scarified by their nibbling. This observation may explain why *R. dominica* in nature is considered to be a 'secondary' pest of grain, coming in only after the grain has been damaged by one of the 'primary' pests e.g., *Calandra oryzae* (Davidson 1941).

A female of the European rabbit *Oryctolagus cuniculus*, forced by social pressures to live a solitary outcast from a social group, will dig a shallow, single-passaged burrow to have her litter in; and she will seek for herself improvised shelter under bushes or in a hollow log. Neither this sort of shelter for the adult nor the shallow burrow for the nest compares favourably with the secure accommodation that is found in the warrens where live the rabbits that are organized into social groups. The burrows in an old well-worked warren are deep and capacious and have many chambers. It takes many rabbits a long time to make such a good place to live.

The termitaria, formicaria, hives where social insects, termites, ants, bees live, also have a quality that comes only from the collaboration of many individuals.

Birch and Andrewartha (1941) described the circumstances in which virtually all the grasshoppers *Austroicetes cruciata* that were living in an area of several hundred square miles in South Australia were eaten by birds. The circumstances were unusual because this occurred during a severe drought. The grasshoppers have one generation a year. The eggs are laid in soil where the birds do not find them. There is a prolonged egg-stage lasting from November to August. The active stages, nymphs and adults, are present from August through November which is the southern spring. We watched this population of grasshoppers for six years; from 1935 to 1939 there were dense swarms distributed widespread over the whole of the 'grasshopper belt' (Fig. 1.01). During this period the birds ate freely of grasshoppers during three months of each year, but the birds were so few relative to the large number of grasshoppers that only a very small proportion of the grasshoppers were eaten. The birds did not increase in numbers during this period because there was no comparable supply of food for them during the rest of the year. In August 1940 the grasshoppers hatched from eggs in the same large numbers. But there had been a severe drought; there was scarcely any green grass; and most of the grasshoppers died from starvation. By October the population which had been large was reduced to a series of small remnants, clustered in local situations that still harboured a little grass because they were a little moister than the rest of the country. The birds sought out the grasshoppers and ate them. Because the number of grasshoppers were few relative to the birds, they were eaten almost to the last one. During the period when the grasshoppers were abundant the chance that any one would be eaten by a bird was small; but during their time of scarcity this risk increased almost to a certainty.

Colonies of the guanay *Phalacrocorax bouganvillii* that are below a certain size do not multiply. It has been suggested that in small colonies the ratio of circumference to area is large and this increases the proportion of chicks that are taken by predators (Hutchinson and Deevey, 1949). Herds of less than about 15 antelopes rarely persist because once the herd has become as small as this it ceases to fight off wolves and coyotes as an organized herd (Leopold, 1933). These are just several examples of the general principle that an animal in a small population may be more likely to be eaten by a 'facultative' predator than one in a large population.

Animals in a large population may have an advantage over those in small ones with respect to temperature, moisture and other components of weather. For example, Cragg (1955) found that a newly-hatched larva of the sheep blow-fly *Lucilia sericata* had little chance of surviving on the skin of a sheep unless the relative humidity remained above 90% for several hours while the maggot penetrated the skin and established a moist place for itself to live. Except in very wet weather scarcely any maggots survived among those that came from egg-masses that had been laid singly on the sheep. But the survival-rate was higher among those that came from aggregations of several egg-masses laid together because the air was more humid in these larger masses. Barton-Browne (1958) showed that *L. cuprina* has a behaviour that makes it more likely to lay egg-masses clustered than solitary. Pearson (1947) measured the rate at which mice *Mus musculus* consumed oxygen at 26°C. At this temperature a substantial proportion of the respiration of a mouse at rest goes towards maintaining the temperature of the body. Pearson compared the oxygen consumed by four mice when they were huddled together and when they were single. The four mice when they were separated used 1·82 times as much oxygen as the same four when they were huddled together. The rate at which oxygen is consumed is an indication of the influence of temperature on the mice. This result showed that the mice experienced a less severe temperature when they were in a group of four than when they were single.

The honey bee *Apis mellifera*, on a cold day, may flutter vigorously. The metabolic heat from the fluttering of many bees may raise the temperature of the hive appreciably above the ambient temperature. Conversely, on a hot day, bees of the nurse caste may seek from workers water instead of nectar and pollen. They spread the water around the hive, which is cooled as the water evaporates. Both these manœuvres are more likely to be effective in a populous hive (Lindauer, 1961).

Andrewartha and Birch (1954) mentioned several other ways in which the animals in small populations may be at a disadvantage compared with those in dense populations.

8.2 INTERACTIONS IN THE ECOLOGICAL WEB MAY BE IMPORTANT

The ecology of *Thrips imaginis* may serve to illustrate how an interaction in the ecological web may be of fundamental importance in explaining the numbers in a natural population (sections 1.1, 3.11). Davidson and Andrewartha (1948a, b) measured the relationship between certain components of weather and the numbers of thrips, using an equation which implied that weather was chiefly important through its influence on the plants the flowers of which are food for *T. imaginis*. An extended summary of this investigation was given by Andrewartha and Birch (1954).

The extrinsic relative shortage of food for *T. imaginis* is extreme during most of the year; it is ameliorated briefly during the spring of each year, but to a different degree from year to year (Figs. 3.02, 8.01). According to Davidson and Andrewartha it was to be expected that the flowers in which the thrips feed would be more abundant, and would be abundant for a longer period in those years (*a*) when the plants were well grown at the end of winter, ready to burst into flower at the first

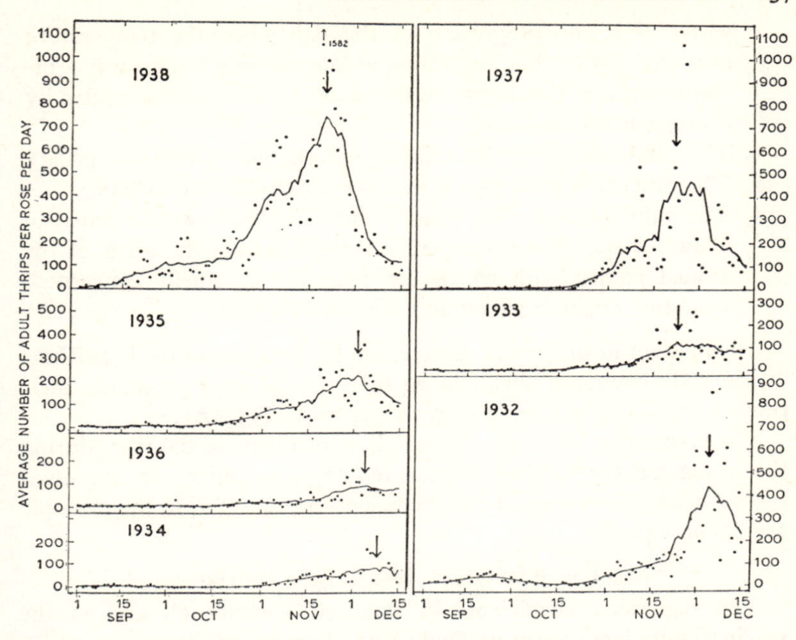

Fig. 8.01. The numbers of *Thrips imaginis* per rose during the spring each year for 7 consecutive years. The points represent daily records; the curve is a 15-point moving average. The arrow indicates the date on which the curve reaches the maximum for each year. Note that the same general trend was repeated during the spring each year, with variations in the height attained by the curve and date on which the maximum was attained. (After Davidson and Andrewartha, 1948*a*.)

sign of spring, (*b*) when the plants received adequate rain to keep them growing during the spring, and (*c*) when the temperature during the spring was high. So they calculated three quantities x_1, x_2 and x_3 from the records of temperature, rainfall and evaporation which seemed likely to express these requirements of the plants.

x_1 the sum of the 'effective' temperature in units of day-degrees from the beginning of the growing season in the autumn until the end of winter on August 31. Temperature was arbitrarily assumed to be effective above 8·9°C and the number of effective day-degrees in a day was estimated as

$$x_1 = P \times \frac{\text{Maximum temperature} - 8{\cdot}9}{2}$$

where P is the proportion of the day when the temperature exceeded 8·9°C. The beginning of the growing season was arbitrarily taken as the day when the aridity of summer was ended by drought-breaking rains.

x_2 The total rainfall for September and October (southern spring).

x_3 The sum of effective day-degrees for September and October.

x_4 The value of x_1 for the year before. This term was included to allow for the chance that, after a good season, the seeds of the annual plants (with which the analysis was chiefly concerned) would be more abundant and more widespread.

The number of thrips was represented by the mean of the logarithms of the daily counts of thrips in 20 roses for 30 days preceding their peak of abundance as indicated by the arrows in Fig. 8.01.

The relationship between the numbers of thrips in the roses during the spring and the weather as measured by x_1, x_2 and x_3 was measured by estimating the values of the coefficients b_1, b_2, b_3, b_4 in the multiple regression equation

$$Y = a + b_1x_1 + b_2x_2 + b_3x_3 + b_4x_4.$$

The regression coefficients in standard measure b^1, and in the original units b are given in Table 8.01. The regression accounted for

Table 8.01. *Regression of log y (numbers of thrips per rose) on* x_1, x_2, x_3, x_4 *(see text) (After Davidson and Andrewartha, 1948b)*

Variate	Regression coefficients		S.E.	t	P
	b^1	b			
x_1	1·224	0·1254	0·0185	6·79	<0·001
x_2	0·848	0·2019	0·0514	3·93	<0·01
x_3	0·226	0·1866	0·1122	1·66	<0·2 >0·1
x_4	0·511	0·0580	0·0185	3·08	<0·02

78% of the variance indicating close agreement, which is also shown visually in Fig. 8.02. From the values of b^1 it can be seen that x_1 was most influential, being about one and half times as influential as x_2. From the values for b it can be calculated (remembering that b is a logarithm and that x_1 was given in units of 100 day-degrees) that on the average the numbers of thrips per rose increased by 41·7% for each increase of 100 day-degrees in x_1.

It is inherent in scientific method that there should be ambiguity

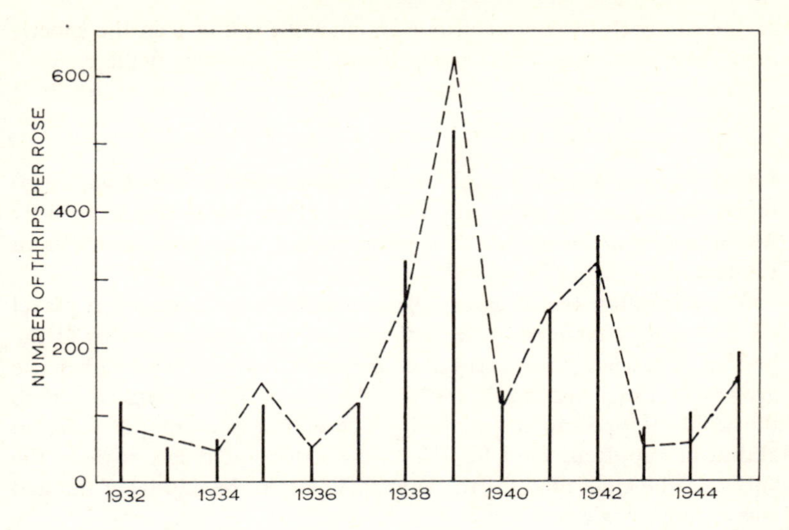

Fig. 8.02. The columns represent the geometric means of the daily counts of thrips per rose. The curve represents the theoretical values for the same quantities calculated from the expression:

$$\log Y = -2{\cdot}390 + 0{\cdot}125x_1 + 0{\cdot}202x_2 + 0{\cdot}187x_3 + 0{\cdot}058x_4$$

and immeasurable (though not unresolvable) doubt about the last step in the investigation when, after a regression, correlation or some other statistical test has failed to disprove a null hypothesis, the scientist must try to equate the terms in his mathematical expression to the causes of the natural phenomena that he has measured. With reference to *Thrips imaginis* it was mentioned in section 3.11 that a part of the influence of the independent variates might be explained by the influence of temperature and moisture, as components of environment, acting directly on the thrips – temperature on the speed of development of all stages and moisture on the survival of pupae. But it seems likely that the most important influence is the indirect one – weather influencing food in the environment of *T. imaginis* and thus ameliorating the relative shortage of food, more some years than others. The significance of b_4, $P < 0{\cdot}02$, may have been caused either by fluctuations in the abundance of seeds left over from the previous year, or by fluctuations in the numbers of thrips left over from the previous year. Either explanation is equally consistent with the mathematics. The second one implies a negative feed-back (see section 9.22). The authors thought that the first one was more likely. They had a lot of experience of *T.*

imaginis to back their opinion but no objective test of it in the experiments that they reported (Davidson and Andrewartha, 1948b).

8.3 INTERACTIONS IN THE ECOLOGICAL WEB MAY RAMIFY

Lloyd (1937, 1941, 1943) and Lloyd, Graham and Reynoldson (1940) found an interesting ramification in the ecological web of the small fly *Metriocnemus hirticollis* which lives in sewage filter-beds in northern England.

A sewage filter-bed is essentially a heap of large water-worn gravel six feet high, contained by vertical walls with a drainage outlet at the bottom. The sewage was sprayed on the surface. In the top foot a dense growth of fungi and algae covered the stones which, together with the solids trapped in it supplied, at certain seasons of the year, an abundant supply of food for the animals that could live so near the surface. The chief food for the animals living in the depths of the bed was a slimy zoogleaa which coated the stones.

During the winter the crust of algae and fungi accumulated on the stones, which became heavily encrusted with it. The thicker this covering became the more likely it was to be loosened by frost and to flake off in large lumps which would sink to the bottom and remain there as food. But when there were many worms and insects feeding in this growth it was more likely to slough off in small fragments, which mostly washed straight through the bed passing out with the effluent, carrying many worms and insects with them. This 'off-loading' not only caused a periodic shortage of food but also depleted the populations of a number of species by washing them away. Off-loading went on continuously, but it reached a peak during spring.

Lloyd *et al.* (1940) listed three species of Oligochaeta, two of Mollusca and seven of Diptera which could always be found in the sewage filter-bed. The different species were favoured by different ranges of temperature. The chance of each one to survive and reproduce was influenced by the activities of one or another of the other species. None was an obligate predator but some would destroy and perhaps eat certain other species incidentally in the course of their feeding.

The small fly *Metriocnemus hirticollis* was one of the least abundant species; the total number trapped in a year, by standard trapping routine, varied from 76 to 846 for the eight years between 1934 and 1941 compared, for example, with 11,000 to 38,000 for a similar species *Spaniotoma minima* which was one of the most abundant species.

Generations of *M. hirticollis* overlapped, and adults were taken in the traps during every month of the year. They were relatively few during autumn and winter and were usually most numerous during April–June. The adults of *M. hirticollis* require to form a swarm in order to mate. The females return to the filter-bed to lay eggs, laying nearly all their eggs in the top foot. The full life-cycle of *M. hirticollis* required about 42 days at 20°C, 123 days at 10°C; none completed development below 7°C. The temperature of the bed varied between 8°C during winter to 17°C during summer. Consequently the activity of *M. hirticollis* was largely restricted to summer. But the eggs laid during June (when the flights of the adults were usually at their greatest) had ample time to become adults before winter, and one might have expected the adults to be abundant during autumn. In fact, they were nearly always scarce at this time. Lloyd (1943) attributed their low numbers at this time to the activities of a related chironomid *M. longitarsus*.

This species thrived at lower temperatures than *M. hirticollis*. A full generation required 26 days at 20°C, 94 days at 10°C and 153 days at 6·5°C. As a consequence, eggs laid early in the autumn emerged as adults during winter; usually, unless the weather was very unfavourable, another large generation was produced before the spring. This led to a large flight during May; the females laid their eggs near the surface and the larvae were well established there by June when *M. hirticollis* was ready to lay its eggs in the same place. The already well-grown larvae of *M. longitarsus*, not predatory by nature, did not seek out the eggs of *M. hirticollis*, but incidentally, in the course of their feeding, destroyed and ate many eggs and small larvae of *M. hirticollis*. Consequently the death-rate among the juvenile *M. hirticollis* was high and the autumnal flights of this species were usually quite small. This sequence of events happened every year. In some years when *M. longitarsus* happened to be less numerous than usual, the rate of survival among *M. hirticollis* was higher, and the numbers of adults greater; the annual totals of adults caught in the traps varied in the ratio 1 : 13. So it is pertinent to enquire into the causes of fluctuations in the numbers of *M. longitarsus*.

Lloyd (1943) recognized two important causes for the fluctuations in the numbers of *M. longitarsus*. The most important was the weather during June when, as a result of large flights during May, there were many eggs and young larvae near the surface. Dry weather during June allowed the surface to dry out and caused the deaths of many eggs and young larvae. On the other hand, moist weather during June permitted a high survival rate and ensured a dense population of well-grown larvae in the filter-bed when *M. hirticollis* was laying its eggs.

The other cause for unusually low numbers of *M. longitarsus* was a warm winter. Since this species was known to be able to develop at moderately low temperatures the opposite might have been expected. Lloyd explained the anomaly by reference to the activities of the worm *Lumbricillus lineatus*. This species was most active during winter; a generation was completed in 110 days at 10°C and 170 days at 7°C. During a warm winter the worms might complete one generation during winter and start another before spring; but when the winter was cold the first generation might not be completed until quite late in the spring. Since the feeding of the worms in the crust of fungi and algae was the chief cause of off-loading, and since this depended largely on the concentration of worms in the surface foot or so of the bed it is clear that off-loading would begin much earlier after a warm winter. After a warm winter the stocks of food for *M. longitarsus* were depleted much earlier than usual and the larvae themselves were at much greater risk of being washed out of the bed. The fluctuations in the numbers of *M. longitarsus* that could be associated with this cause were considerable: the numbers of adults trapped during May varied from 20 in 1938 to 860 in 1941.

Lloyd (1943) found no evidence that *M. longitarsus* needed or benefited from the eggs or larvae of *M. hirticollis* in its diet: it was in no realistic sense a predator. Because it was not a predator the numbers of *M. longitarsus* were not in any way influenced by the numbers of *M. hirticollis*. As a malentity, *M. longitarsus* pressed heavily on *M. hirticollis* keeping its numbers low; but there was no recognizable feedback. The causes of the fluctuations in the activity of this malentity had to be sought more remotely in the recesses of the ecological web – weather influencing the activity of *Lumbricillus* which influenced the supply of food for *Metriocnemus longitarsus* and also caused a malentity in its environment – a crumbly medium which led to off-loading.

Apples in commercial orchards are now almost universally grafted on resistant root-stock which makes the roots an unsuitable place for the aphid *Eriosoma*. The predator *Aphelinus* has been spread to most places where *Eriosoma* occurs. So *Eriosoma* is now a rare species almost everywhere (section 5.11). But occasionally one comes across seedling apples, growing on their own roots. In these circumstances *Eriosoma* may flourish on the roots but not on the aerial parts if *Aphelinus* is also present. In the absence of *Aphelinus* both roots and branches would be favourable places for *Eriosoma*.

The European starling was introduced into North American about 1890. It is now widespread and abundant. On the whole it lives in different sorts of places from the bluebird *Sialis sialis* and the flicker *Colaptes auratus* which still have a large part of their populations living away from towns; but neither the bluebird nor the flicker live in towns as much as they used to do before the starling came.

8.4 ECOLOGICAL BARRIERS AS A PART OF THE ECOLOGICAL WEB

The theme of Howard's (1931) book *The insect menace* was that the modern agricultural practice of growing crops of a single species, densely crowded in contiguous farms over large areas, was likely dramatically to alleviate the relative shortage of food that might otherwise have been experienced by many insect pests, or potential pests, when the same species of crops were grown in small patches intermingled with other species, as in the older agricultural practices. The same theme was developed by Voûte (1946) in a number of papers comparing the damage done by insect pests in artificial forests of a single species with the damage done in natural forests where a number of different species are interspersed.

Fisher and Ford (1947) studied a colony of the moth *Panaxia dominula* which occupied about 20 acres of fenlike marsh near Oxford. The characteristic vegetation included *Symphytum officinale* which was the chief food of the larvae. The marsh was bounded by woodland and agricultural land, into which the moths seemed never to penetrate notwithstanding that they flew strongly. Entomologists have collected in this vicinity for many years and their testimony (reliable because *Panaxia* is a large brightly coloured day-flying species) confirms that *Panaxia* is not to be found straying beyond the confines of the specialized area of vegetation where the colony lives.

On a much grander scale it is known that forests of *Pinus* in North America used to be free from the sawfly *Neodiprion sertifer* (a serious pest of pines in Europe) until it was accidentally introduced (across the ecological barrier of the Atlantic ocean) some time before 1925 (section 5.22). Now it is widespread and a serious pest in North America.

These few examples may serve to make the point that an ecological barrier either small or large that reduces the compactness of the distribution of a resource may influence its accessibility which is the essence of a relative shortage.

9 *Theory: the numbers of animals in natural populations*

9.0 INTRODUCTION

Animals are usually distributed patchily. Even when one arbitrarily selects what seems to be a uniform small area one finds that the animals are not distributed uniformly or even randomly over it; they tend to occur in 'clumps'. This may be partly explained by accidents in the history of the population, by the behaviour of the animals (e.g. gregariousness) or by the heterogeneity of the terrain (section 10.3). The last-named cause becomes increasingly important as the area becomes large. In large areas, of the order that we usually choose for a practical ecological investigation, irregularities of terrain are pronounced and this, in itself, leads to 'colonial' or 'clumped' distributions of the animals in the area (Elton, 1949). For a description of an extremely colonial distribution see the passages quoted from Cockerell in section 3.12 and Flanders in section 10.44. See also section 10.45.

A general theory about the numbers of animals in natural populations, if it is to be realistic, must take into account the uneven distribution of the animals. Andrewartha and Birch (1954) suggested that one way to do this was to regard the whole population as the sum of a number of local populations. They examined the conditions of commonness and rareness in local populations as a first step and then extended the principles to cover the larger population comprising many local populations. I have followed the same method in this book and sections 9.1 and 9.2 are based largely on Chapter 14 of Andrewartha and Birch (1954).

9.01 *The meanings of 'common' and 'rare'*

If we say that a species is *rare* or *common* we imply that individuals of this species are few or many relative to some other quantity that we can measure. We might, for example, relate them to some unit of area and judge them rare if they are, say, fewer than one (or one million) per square mile. This is the commonsense meaning for the hunter and the fisherman, and it is the meaning that I shall use in section 9.2.

Another way of considering commonness and rareness is to relate the number of individuals in the population to the quantities of necessary resources, food, nesting sites, etc., in the area that it inhabits. This may be the commonsense meaning for a farmer who may not be interested in knowing how many caterpillars there are per square mile; it may be more important for him to know whether they are numerous enough to eat much or little of his crop. This is the meaning that I use in section 9.1.

9.1 THE CONDITIONS OF 'COMMONNESS' AND 'RARENESS' IN LOCAL POPULATIONS

9.11 *The conditions of commonness in local populations*

The codlin moth that were mentioned in section 3.211 were common relative to their stocks of food – so common in fact that there was a caterpillar in almost every apple. But they did not, by virtue of their presence, reduce the size of the next year's crop. In other words, although they multiplied up to the limit of their stock of food they did not, by their numbers, reduce the amount of food for the next generation. A similar condition was described by Kluijver (1951) for a population of great tits *Parus major*. The tits were living in an area of mixed woodland where most of the trees were young with few holes suitable for nests; the birds consistently used up all the nesting sites year after year.

This condition of commonness in a local population may be represented by the curve and the symbols in Fig. 9.01. This diagram is generally true of any local population whose numbers are determined by the stock of some non-expendable resource – non-expendable in the sense that the amount available to the next generation is independent of the amount used by the present generation (Table 3.01 classifications Ba and Bb).

The local populations of *Cactoblastis* on prickly pear in Queensland (sections 3.12, 3.2) represent the converse of this condition. The first ones to arrive at a place where prickly pear is growing multiply rapidly. But the food, being a growing plant, tends to be reduced in relation to the number of *Cactoblastis* feeding on it so there is less food for each succeeding generation; eventually there comes a time when there is none left for the next generation; and the population in that locality dies out – but not, as a rule, before it has given rise to emigrants, some of which may find a fresh supply of prickly pear and start the cycle all over again in a new place. This sequence of events is illustrated by the diagram in Figure 9.02. This diagram is generally true of all local populations whose numbers depend on the amount of some diminishing or expendable resource – expendable in the sense that the more that is used by one generation the less there will be for the next.

9.12 *The conditions of rareness in local populations*

The woolly aphis which I discussed in section 5.11 never consumes more than a fraction of the food in its area. During the summer no local colony of *Eriosoma* lasts for long without being discovered by the predator *Aphelinus*; and shortly after it has been discovered it is annihilated – but usually not before a few emigrants have gone off with a chance of founding a new colony on a neighbouring tree. Thus the aphid remains rare relative to the large amounts of easily accessible food that surround it. I have illustrated this sequence of events in Fig. 9.03 with the series of symbols that ends with a circle that is empty except for the cross inside it; this symbol indicates that there used to be a colony of animals here until recently.

As winter approaches, the lower temperatures at first hamper the development of *Aphelinus* relatively more than that of its prey. Colonies of *Eriosoma* that are discovered by *Aphelinus* late in summer may not be annihilated before both species become dormant for the winter. With rising temperatures in the spring *Eriosoma* is able to multiply while *Aphelinus* still remains dormant. This sequence of events is represented in Fig. 9.03 by the series that ends with a circle with a small segment shaded; this symbol indicates that a small remnant of the population persists and becomes the nucleus which multiplies again when circumstances permit. The overall size of the circles remains unchanged because the animals are few relative to the amount of food; so they make no appreciable difference to the amount that is present for succeeding generations.

The diagram in Fig. 9.03 with its two alternative endings has a wide application. It can be taken quite generally to represent the broad and important condition of rareness in which any component of environment keeps the population rare relative to its stock of food, nesting sites, etc. I have discussed this diagram in relation to a population that was kept in check by a predator, but the check might equally well have been exercised by weather, the recurrent use of an insecticide or any other component of environment.

I mentioned in section 8.3 that when *Eriosoma* colonizes a seedling apple that is growing on its own roots the aphis can thrive on the roots free from attack by *Aphelinus*. But those that colonize the branches are almost certain to be destroyed by the predator quite soon. I have represented such a situation in Fig. 9.04. This diagram may be taken generally to represent any population in which a more or less constant proportion is secure against a risk that is likely to destroy the remainder. With *Eriosoma* those that were not sheltered underground were likely to be killed by a predator. Territorial vertebrates for which the possession of a territory secures the occupant against the risks of predation are like *Eriosoma*. But the diagram has a much wider connotation than this. It also covers any extrinsic relative shortage of resource such as was discussed in section 3.11; the risk for those animals, not lucky enough to find food (if that were the resource in question) would be starvation.

9.13 *The way in which weather may keep a local population rare relative to food and other resources*

In regions with temperate climate, and perhaps in most other climates as well, many species have well defined breeding seasons and most species experience only a limited period that permits multiplication. For the rest of the year deaths exceed births and the numbers decrease. Consequently it is characteristic of populations whose numbers are determined largely by weather, that they should fluctuate more or less in step with the seasons (Fig. 3.02). And what is true of seasons may also be true of longer periods. Runs of favourable years may be followed by runs of less favourable ones. Thus the average density that is attained over any period of years depends on the relative rate of increase and decrease during the favourable and unfavourable periods, and the relative duration of the favourable and unfavourable periods. I have illustrated this principle in Fig. 9.05 (see also Fig. 1.03). In order to simplify the exposition I have drawn three pairs of curves so that rate of increase, extent of decrease and duration of favourable period may be compared independently. But in nature these three comparisons may not be independent of each other. The difference between two places at the same time or the same place on separate occasions may be compounded of all three comparisons.

In Fig. 9.05 the abscissae are in units of time – days, years or generations. The ordinates for each curve are numbers of animals; *K* represents the number that the area would support if the total resources of food were used up. The number indicated by Y_0 is the number left when the unfavourable period ends and is also the nucleus for multiplication when the next favourable period begins. The rate of increase that pertains during the favourable period is called *r*. Suppose that the three pairs of curves represent three ways that weather may influence the numbers of animals in a local population. The curves on the left relate to a place where the weather is favourable, and those on the right relate to a place where the weather is severe. In the top pair the two curves start from the same level at the beginning of the favourable period (Y_0 is constant); and they rise at the same rate because *r* is constant. But the curve for area A rises farther because the favourable period lasts longer. With the middle pair the two curves rise at the same rate (*r* is constant); they continue rising for the same time (favourable period, *t* is constant); but the curve for area A rises farther than that for B because it started from a higher level (Y_0 is greater for A). This means that for some reason or another the unfavourable period was less prolonged or less severe in A than in B so that the population in A was still relatively large when the weather changed and allowed the animals to start increasing again. With the bottom pair the two curves start from the same level (Y_0 is constant); they continue rising for the same interval of time (favourable period, *t*, is constant); but the curve in A rises farther because it is steeper (rate of increase, *r*, is greater for A). Taking all three curves into account and compounding them, one can easily see that the animals would, on the average, be more numerous in area A than in B. The same idea is expressed differently in Figs. 1.02 and 1.03.

Figure 9.05 may also be considered to represent the condition of rareness that prevails in a population of insects that is 'controlled' by insecticides. The first pair of curves compares the results of frequent and infrequent applications of the insecticides. The second pair of curves compares the result of spraying thoroughly with a good insecticide with the result of spraying less thoroughly or with an inferior insecticide. The third pair of curves illustrate what might happen if the same treatment were used against two populations of the pest one of which was able to multiply more rapidly than the other.

9.2 THE CONDITIONS OF 'COMMONNESS' OR 'RARENESS' IN NATURAL POPULATIONS

When we come to consider the whole population in an area that is large enough to support many local populations it is necessary to broaden the meaning of 'common' and 'rare' to include the idea of so many per unit area (section 9.01 – first meaning). There are four points to be considered. (*a*) What are the conditions of commonness or rareness in the local populations? (*b*) Are the local populations in step? Or are different ones at different stages of development at the same time – some increasing while others are decreasing and so on? (*c*) How many local populations are there in a unit of area and how large are the individual local populations? (*d*) How relevant is the principle of relative shortage of food (section 3.1)?

With these four points in mind I have constructed a series of diagrams which are intended to illustrate the conditions of commonness or rareness in natural populations, i.e. populations that occupy substantial areas of the dimensions that ecologists usually choose to study.

In Fig. 9.06 each symbol represents the condition, at one moment of time, in a local population of the sort that can be represented by Fig. 9.01. Area A may be compared with area B. In both areas the local populations are similar (in that the animals are common relative to the stock of some unexpendable resource) but there are more of them in A than in B. Consequently the population in the whole area is more numerous in A than in B.

In Fig. 9.07 each symbol represents the condition, at one moment of time, of a local population of the sort that can be represented by Fig. 9.02. It is characteristic of this condition that the local populations are likely to be at different stages of development at the same time (section 3.12), and I have allowed for this in drawing the diagram. I have also allowed for differences in the frequency of local accumulations of food and in the dispersiveness and ability to search, of the different sorts of feeding animals. There are, therefore, three comparisons that may be made within Fig. 9.07. Both the species that occur in area A and the one that is found in areas B and C multiply, in local populations, to the limit of the food supply and then die out from an absolute shortage of food (section 3.2). The species in area A is highly dispersive; the one in areas B and C is less so. In considering the whole area it would seem clear that the animals in C would experience a severe shortage of food in the relative sense; and in a different way the same is true of those living in A also. The animals in A are so dispersive that very little food goes undiscovered for very long; consequently the food becomes scarce and sparsely distributed, and the animals come to experience a severe relative shortage of food.

Figure 9.08 is made up from symbols taken from Fig. 9.03 in order

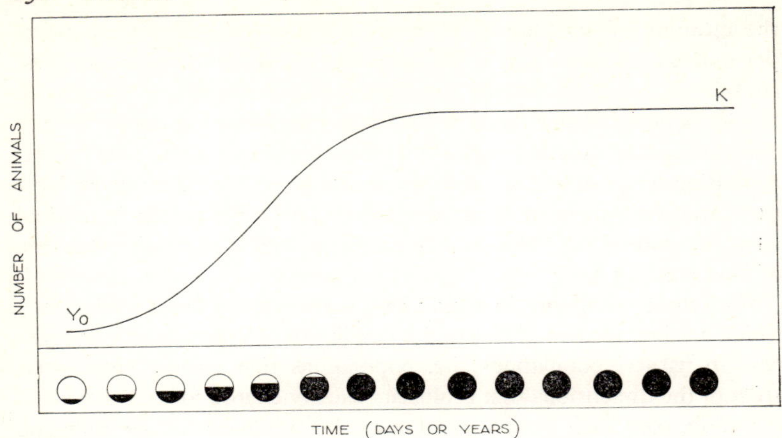

Fig. 9.01. The growth of a 'local population' (i.e. the population in a 'locality') whose numbers are limited by the stock of some non-expendable resource, such as nesting sites. The initial numbers (i.e. the number of immigrants who originally colonized the locality) are represented by Y_0; the maximal numbers that the resources of the locality will support are represented by K. The circles repeat the information given by the curve. The proportion of the circle shaded represents the number at a specific time as a proportion of the maximum.

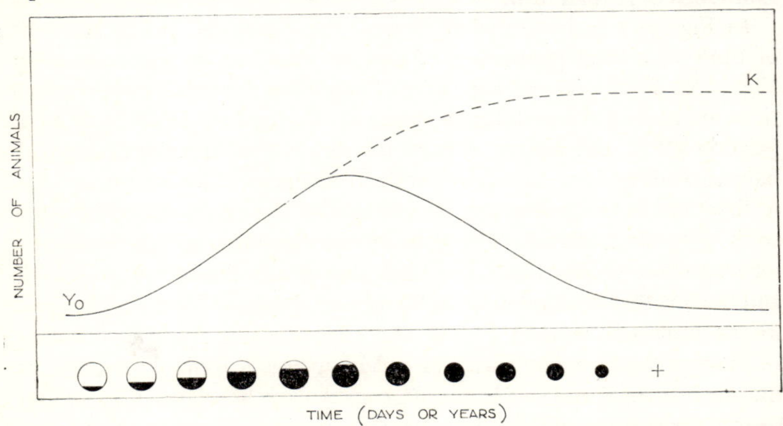

Fig. 9.02. The solid line represents the growth of a local population whose numbers are limited by a diminishing resource, such as food, which becomes less, the more animals there are feeding on it. (The broken line indicates the numbers that the population might attain if the resource were non-expendable, as in Fig. 9.01.) The symbols have the same meaning as in Fig. 9.01; the circles grow smaller because the plants (or population of prey) become fewer and eventually die out because there are too many animals eating them. The cross indicates that this is a place where a local population recently became extinct (see Figs. 9.07 and 9.08). (After Andrewartha and Birch, 1954.)

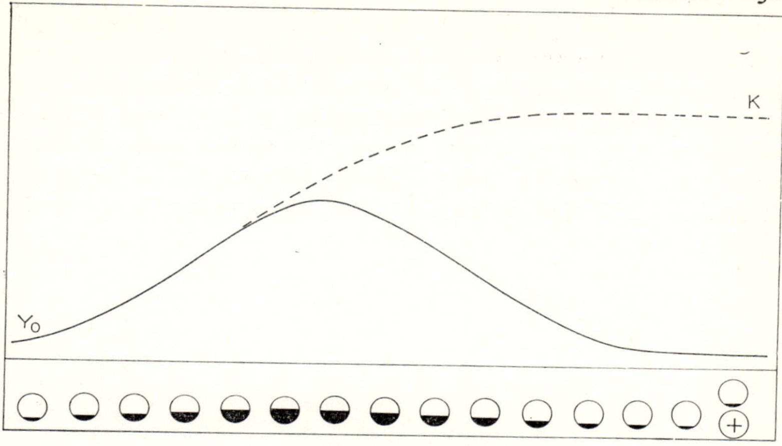

Fig. 9.03. The solid line represents the growth of a local population that never uses up all its resources of food, etc., in the locality because its numbers are kept by weather, predators, shortage of food in the relative sense (section 3.1), or some other environmental component, at a level well below that which the resources of the locality could support. The broken line and the symbols have the same meaning as in Figs. 9.01 and 9.02. The two symbols at the end indicate alternative endings to the cycle of growth of the local population: the population may be extinguished in this locality (*circle with a cross in it*) or a remnant may persist, perhaps to increase again when the circumstances change (*circle with a remnant of shading in it*). The circles do not become smaller because the resource, although it is of the sort that would diminish if many animals used it, nevertheless does not diminish appreciably because the animals do not become numerous enough (e.g. *Icerya* section 3.12). (After Andrewartha and Birch, 1954.)

to illustrate the way that a predator may influence the abundance of its prey. The condition of the population in A may be compared with the description of *Saissetia* (as prey for *Rhizobius* in California) and the condition of the population in area B may be compared with the description of *Icerya* in section 5.11.

The two areas A and B support populations of the same species of herbivore. These herbivores are not numerous enough to reduce their stocks of food appreciably, so there are the same number of circles drawn in both areas. In area B there is an active predator with powers of dispersal and rate of increase which match those of the prey; in area A these predators are replaced by another species with equivalent rate of increase but with inferior dispersiveness. (That is, it annihilates local populations just as effectively but it finds them less frequently.) In B many localities that are quite favourable are empty; some have become empty only recently (circle with a cross in it), and in most of those that are occupied the numbers are low. Either the prey have only recently arrived at the place and have not yet had time to become numerous, or else they have been there longer and are in the process of being exterminated by the predators. Contrast this with A where relatively more places are occupied and the local populations are, on the whole, larger. There are relatively few places from which the prey have recently

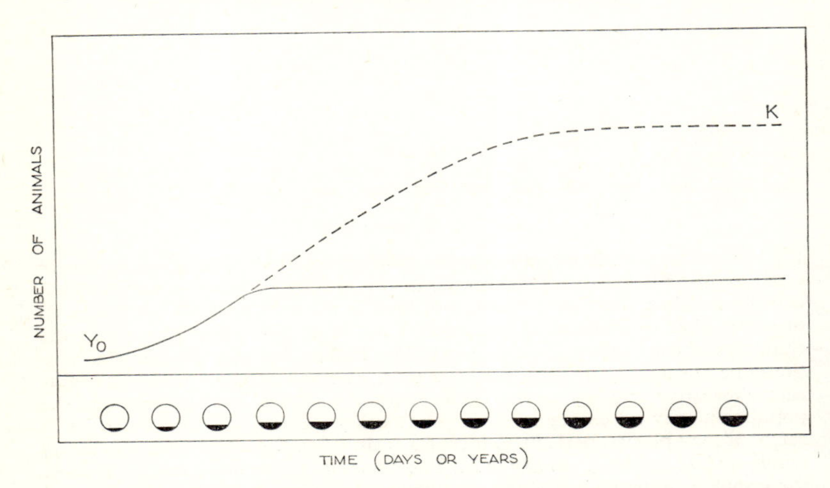

Fig. 9.04. This diagram has two meanings: (*i*) The solid curve represents the growth of a local population in which a more or less constant number of the individuals are sheltered from some risk which is likely to destroy all those that are not so sheltered. This number is not enough to use up more than a small proportion of the resources of food, etc., in the locality. (*ii*) The solid curve represents the numbers that, on the average, manage to get enough of a resource in the face of an extrinsic relative shortage as for example *Glossina* or *Ixodes* (section 3.11). In both meanings the dotted line represents the number of animals that might be supported if all the resource were accessible.

been exterminated. The prey in area A have the advantage of being more dispersive than their predators and this allows them a longer interval for multiplication in each new place that they colonize.

The way that an extrinsic relative shortage of a resource, or weather influences the density of the population may be illustrated in Fig. 9.09. It is based on Figs. 9.04 and 9.05. In Fig. 9.09 there are the same number of places where the animals may live in A and B, but the weather is more favourable in A than B, so the animals, in each local population, are more common, relative to their stocks of food, etc., in A than in B;

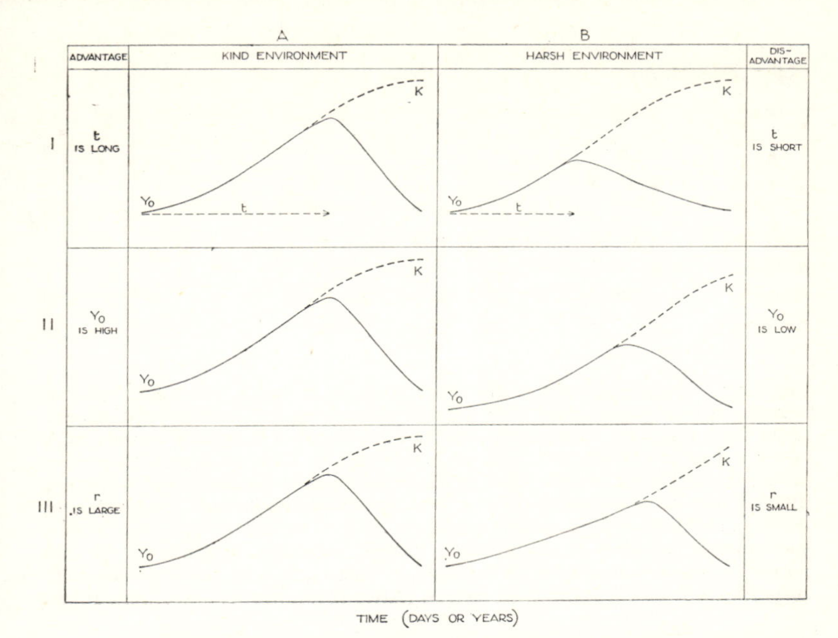

Fig. 9.05. Three ways in which weather may influence the average numbers of animals in a locality. The numbers in the locality are increasing during spells of favourable weather and decreasing during spells of unfavourable weather. Two areas A and B are compared with respect to three qualities. Quality I is related to the duration of the favourable period; quality II is related to the severity of the unfavourable period; and quality III is related to the favourableness of the favourable period. The numbers that would be attained if all the resources of food, etc., were made use of are indicated by K; the numbers to which the population declines during the unfavourable period are represented by Y_0. For further explanation see text. (After Andrewartha and Birch, 1954.)

Fig. 9.06. The populations of large areas are made up of local populations. In the two areas which are compared in this diagram the resources of every locality are fully used up (as in Fig. 9.01) but there are more favourable localities in area A than in area B. (After Andrewartha and Birch, 1954.)

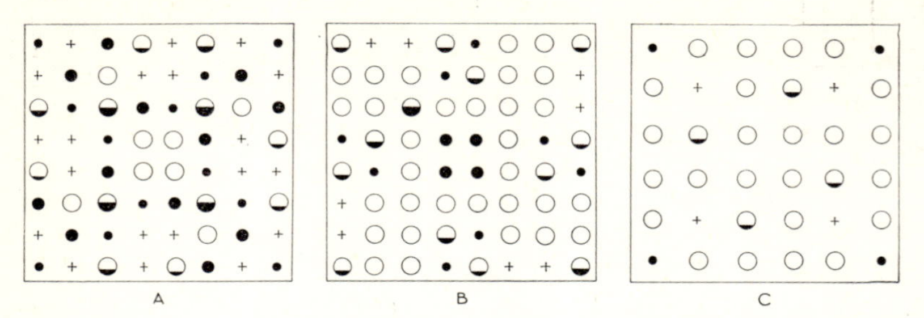

Fig. 9.07. The populations of the large areas A, B and C are made up of a number of local populations. The symbols describing the local populations are taken from Fig. 9.02. Favourable localities are distributed equally densely in A and B but more sparsely in C. Area A is occupied by a species with high powers of dispersal; B and C are occupied by a species which is similar except that it has inferior powers of dispersal. In the local situations the animals experience an absolute shortage of food. In all three areas A, B, C taking the areas as a whole, the animals experience an intrinsic relative shortage, accentuated by the naturally sparse distribution of the food in area C, and by the naturally high dispersiveness of the animals in area A. (After Andrewartha and Birch, 1954.)

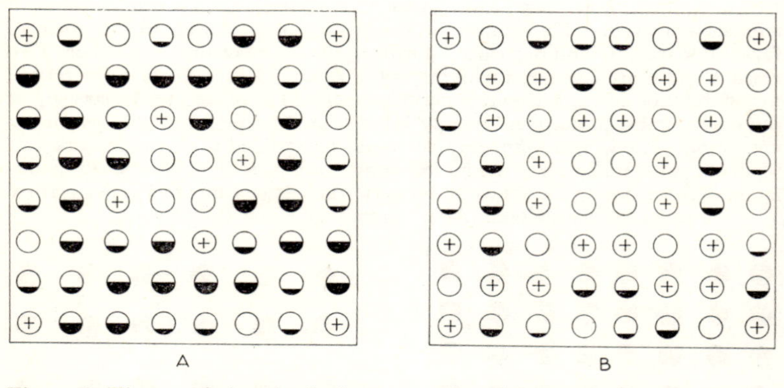

Fig. 9.08. The populations in the large areas A and B are made up of a number of local populations. The symbols describing the local populations are taken from Fig. 9.03. Favourable localities are distributed equally in the two areas. In area B there is an active predator whose powers of dispersal and multiplication match those of the prey. This predator is absent from A where its place is taken by one which is more sluggish. Relatively more of the favourable places are occupied in A than in B, and the local populations are, on the average, larger. Altogether the animal is more abundant in A than in B. (After Andrewartha and Birch, 1954.)

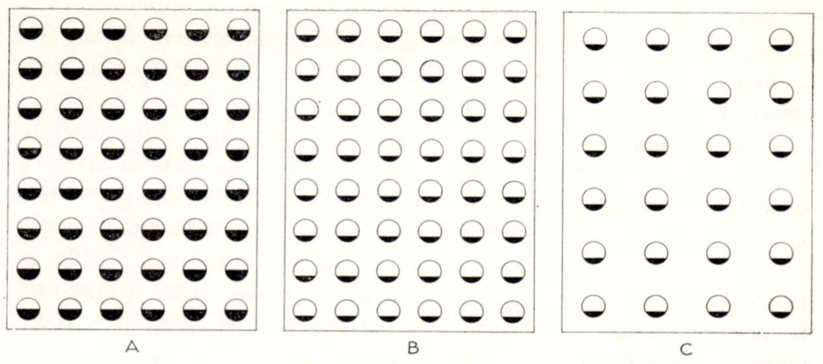

Fig. 9.09. This diagram has two meanings. (*i*) The populations in the large areas A, B C are made up of many local populations of the sort illustrated in Fig. 9.04. In A either a greater proportion of the population is sheltered than in B or C or the extrinsic relative shortage is less severe than in B or C. (*ii*) The local populations are influenced directly by weather as in Fig. 9.05; the climate in A is more favourable than in B or C. In both meanings the number of favourable localities is greater in A and B than in C. (After Andrewartha and Birch, 1954.)

consequently the population in the whole area is larger in A than in B. Area C is inferior to B with respect to the number of places and C is inferior to A with respect to both the number of places and the weather. So the animals are rare in C relative to A, in both meanings of the word (section 9.01). Alternatively the extrinsic relative shortage is less severe in A than in B or C. See the legend under Fig. 9.09.

9.21 *General conclusions*

Figures 9.03, 9.04, 9.05B, 9.07B, C, 9.08B and 9.09B, C, illustrate populations in which animals are rare relative to their stock of food (or other resource). This is true of the whole population in Fig. 9.07C even though the local populations consume all their food. In Fig. 9.07 (not in the local populations but in the area considered as a whole) the number of animals is determined chiefly by a relative shortage of food (sections 3.0, 3.1). In every other instance where the animals are rare relative to their stock of food (or other resource) the number in the population is determined by a shortage of time. This is immediately obvious from inspection of Fig. 9.05. But inspection of Figs. 9.03 and 9.08 also leads to the same conclusion: each local population is likely to be exterminated soon after it has been found by the predators; how numerous the animals become in each local population depends largely on the interval of time that elapses between the founding of the colony and its discovery by the predators; when this interval is long the animals become relatively abundant, when it is short they remain relatively scarce.

Figures 9.01, 9.02, 9.05A, 9.06, 9.07A, 9.08A and 9.09A illustrate populations in which the animals are abundant relative to their stock of food (or other resource). When, as in Figs. 9.01 and 9.06, the resource is not expendable, the number in the population depends quite simply on the total amount of resource. When the resource is expendable, as in Figs. 9.02 and 9.07, the number in the population may still be said to depend on the amount of resource but the relationship is more subtle because the amount of resource also depends on the number of animals: that is, the relationship is to some extent reciprocal. In the local populations the numbers can be explained in terms of the absolute shortage of food (section 3.2); in the natural population, comprising many local populations, the number of animals depends on the absolute amount of resource and its accessibility (shortage in the relative sense, section 3.1).

I conclude that the numbers of animals in natural populations may be explained by either (*i*) a shortage of time or (*ii*) a relative shortage of a resource.

When the shortage of time relates to the time that a local population might be free from discovery by a predator, and when the relative shortage is intrinsic a negative feed-back to density might develop. So it is important to inquire into the conditions for negative feed-back in natural populations (section 9.221).

9.22 *Negative feed-back to density*

So much has been written about negative feed-back to density in natural populations that to summarize it in depth is beyond the scope of this book. Smith (1935) pointed out that, as a consequence of the natural variability in the quality of a resource, some individuals might do better than others: so the proportion doing badly might be larger in a dense population. Andrewartha and Birch (1954) suggested that a large population, being more variable genetically, might contain a larger proportion of weakly individuals relative to any component of environment. Chitty (1967; section 5.3) built his general theory of self-regulatory mechanisms in natural populations around an extension of this idea. Other authors, notably Nicholson (1954) and Solomon (1957), following the tradition of Howard and Fiske (1911), emphasized the reaction between the activity of a particular component of environment and the density of the population (Fig. 9.10 cells B and E). In their classical paper Howard and Fiske emphasized 'parasitism' (predation in the usage of this book):

> . . . it is necessary that among factors which work together in restricting the multiplication of the species [caterpillars of *Porthetria dispar* and *Euproctis chrysorrhoea* which defoliate forest trees] there shall be at least one, if not more, which is here termed facultative (for want of a better name) and which, by exerting a restraining influence which is more effective when other conditions favour undue increase, serves to prevent it. . . . Very few other factors [influencing the survival of the caterpillars] however closely they may be scrutinised, will be found to fall into the class with parasitism, which in the majority of insects, though not in all, is truly 'facultative'. . . . A natural balance can only be maintained through the operation of facultative agencies which effect the destruction of a greater proportionate number of individuals as the insect in question increases in abundance.

Nicholson placed much emphasis on 'competition' for resources or by predators (see Andrewartha and Birch, 1954, p. 22). It will be seen (section 9.221) that there are many interactions that generate negative feed-back either in the broad usage of Howard and Fiske or in the restricted usage of Nicholson and Solomon. Indeed, there is general agreement among ecologists that at least some measure of negative feed-back occurs widely, perhaps universally in natural populations. But there has also been much controversy.

Two streams can be recognized in this controversy, one fruitless and boring, the other fruitful and exciting. On the one hand, it is argued (strictly in the tradition of Howard and Fiske) that only density-governed reactions can maintain a natural balance in a population; but the density-governed reaction is usually defined more restrictively by modern authors than by Howard and Fiske (Nicholson, 1954; Solomon, 1957). In so far as much of the argument in this stream of the controversy either states a conclusion that seems not to need to be tested empirically because it has been deduced by 'logical argument based on certain irrefutable facts' (Nicholson, 1958, p. 114), or restates the truism that animals need resources and resources are not infinite, it does not concern us. It is doubtful whether this theory could be tested empirically because the only particular statements that are consistent with it are statements about 'balance' (Nicholson, 1954, p. 60) or 'ultimate upper or lower limits' (Elton, 1949) which cannot be translated into the design of any practicable experiment. So this stream of the controversy is fruitless because, as Popper (1959) pointed out, propositions that, either for logical or technical reasons, cannot be tested empirically, or propositions that do not need to be tested empirically because they are truisms, or logical deductions from truisms ('irrefutable facts'), are outside the realm of science.

On the other hand, there is a fruitful and exciting discussion to be had about the actual mechanisms of negative feed-back and how important they might be in explaining, not 'balance' or 'ultimate limits' which are intangible and untestable, but the actual numbers of animals that might be counted or estimated in a natural population during a finite period.

It was shown in section 8.2 that the numbers of *Thrips imaginis* in a natural population, estimated by daily counts for 14 years, could be explained (78% of the variance) by four terms of a multiple regression, three of which represented environmental components that unequivocally were not associated with density-dependent reactions; the fourth was consistent with an explanation that included a density-dependent interaction, but the authors preferred another explanation which did not invoke density-dependence. Also it is difficult to see how the idea of negative feed-back might enter into the explanation of the numbers of *Metriocnemus hirticollis* counted during eight years by Lloyd (section 8.3). I have made a close study of a number of papers by Frith (1959a, b, c, d, 1962, 1963) on the ecology of the grey teal *Anas gibberifrons* in Australia. I have also studied a long series of papers on the ecology of the European rabbit in Australia* (see references in section 3.212). I concluded that a full explanation of the numbers of both these species in nature could be made without invoking the idea of density-dependent reactions. I think that when the ecology of more species are known well density-dependent reactions will be found to be important in relatively few of them. This is not to deny that density-dependent reactions may occur in many ecologies and may prove to be important in some.

9.221 The conditions of negative feed-back in natural populations

Figure 9.10 is a two-way table. The columns distinguish between components of environment according to whether the influence of the environment on the animal is positive or negative. The graphs that define the columns have been explained in the introductions to chapters

* I hope to publish ecological case-histories elsewhere.

3–7. The influence of the component of environment is arbitrarily said to be positive when, as in the case of resources, mates and weather (on the lower side of the optimum), the animal's chance to survive and reproduce increases with increasing activity of the resource. Conversely the influence of the component of environment is negative when, as in the case of predators, aggressors, malentities and weather (above the optimum), the animal's chance to survive and reproduce decreases with increasing activity of the environmental component.

The rows in Fig. 9.10 distinguish between components of environment according to whether they respond positively, negatively or not at all to the density of the population. Because this response depends on circumstances the same component of environment may appear in more than one row, to distinguish, for example, between an intrinsic and an extrinsic relative shortage of a resource (sections 3.11, 3.12), or an obligate and facultative predator (section 5.1). Because this response takes time the density of the population (N_t) refers to time t and the response in the activity of the environmental component (A_{t+1}) refers to a period one unit of time later. It may be convenient to consider a generation or a breeding season as equal to one unit of time. There is no time-dimension in the curves at the head of the columns so the two sets of curves cannot be related in time. Each set of curves is independently diagnostic of the interactions that are relevant to the classification in the columns and the rows respectively.

The derivation of the curves that define the rows may be studied, by following the references given in Table 9.01, back to the appropriate sections. For example, the curve that defines row III was discussed with reference to resources in section 3.12 and in columns A(*a*) and A(*b*) of Table 3.01; and the curve that defines row I covers aggressors because Chitty's hypothesis states that aggression is selected for in a dense population (section 5.3).

The classification in Fig. 9.10 places ecological reactions into particular cells in the two-way table according to whether the reaction generates a positive feed-back, a negative feed-back or no feed-back at all. Because the reactions between an animal and other individuals of its own kind vary in complexity (sections 3.11, 4.0, 5.3, 8.0) so the reactions classified into different cells in Fig. 9.10 also vary in complexity, some extending further into the ecological web than others. But for this analysis the complexity of the reaction matters less than its outcome.

Reactions that give rise to positive feed-backs are placed in cells A and F. Reactions that give rise to negative feed-backs are placed in the

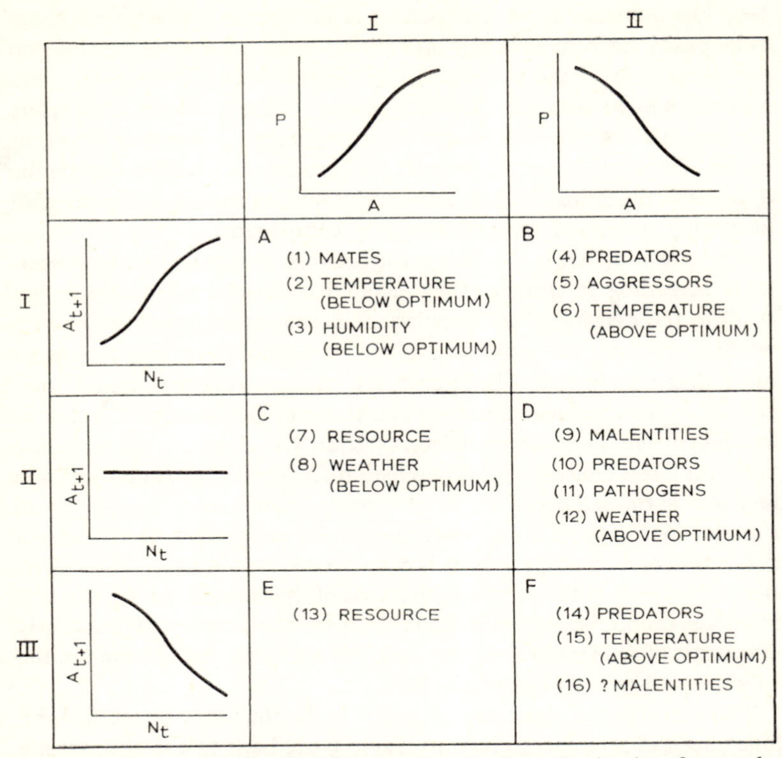

Fig. 9.10. Two-way table classifying reactions between the density of a population and the activity of the environment according to whether the reaction gives rise to a positive feed-back, a negative feed-back or no feed-back at all. Positive feed-backs appear in cells A and F; negative feed-backs in cells B and E and no feed-back at all in cells C and D. Column headings define the influence of any component of environment on an animal's chance to survive and reproduce; Row headings define the reaction of the environmental component to the density of the population. N_t, density of population at time t; A_{t+1}, the activity of the environmental component after one unit of time, i.e. at time $(t + 1)$. Comments and explanation of the entries in the cells appear in table 9.01 (opposite page).

corner cells of the opposite diagonal B and E. Components of environment operating in circumstances that generate no feed-back are placed in the central cells C and D. These events (non-reactions) correspond to the classical 'catastrophic factors' of Howard and Fiske (1911); they have been given a variety of names since then in theories that have followed Howard and Fiske more or less closely ('density-independent

Table 9.01. *Comment on the entries in Fig. 9.10 (opposite page)*

Reaction	Comment	Reference section
(1) Mates	*Ixodes ricinus*; the risk of not meeting a mate in a sparse population, and failure of small colonies to colonize a new area	4.1
(2) Weather	*Apis mellifera*; on a cold day bees flutter and so keep the hive warm.	8.1
(3) Weather	*Lucilia sericata, L. cuprina*; the larvae hatching from a single egg-mass fail to establish on healthy skin because it is too dry but larvae from large aggregations of eggs succeed.	8.1
(4) Predators	*Icerya purchasi* has an obligate predator *Rodolia cardinalis* in its environment.	5.11
(5) Aggressors	*Microtus agrestis*; according to Chitty's hypothesis.	5.3
(6) Weather	*Calandra oryzae*; a dense population in a large bin of wheat causes the temperature of the wheat to rise intolerably.	6.12
(7) Resource	*Glossina morsitans* is the classical example of an extrinsic relative shortage of food.	3.11
(8) Weather	*Drosophila melanogaster*; the typical relationship between temperature and the speed of development of a poikilotherm is shown in Figure 6.02.	6.1
(9) Malentities	Collembola may be trapped in small puddles.	7.1
(10) Predators	*Levuana irridescens* has a facultative predator *Ptychomyia remota* in its environment.	5.11
(11) Pathogens	*Myxoma* is transmitted between rabbits by mosquitoes.	5.21
(12) Weather	Goldfish die if the temperature is too high.	6.13
(13) Resource	*Cactoblastis cactorum*, by its own feeding, causes an intrinsic relative shortage of food.	3.12
(14) Predators	Antelopes; herds of less than 15 may not fight off predators as an organized herd.	8.1
(15) Weather	*Apis mellifera*; on a hot day bees of the nurse caste demand water from workers and spread it around the hive.	8.1
(16) Malentities	Hypothetical malentity which is destroyed by a dense population. It is a logical possiblity but I do not know of an empirical example.	—

factors', Smith, 1935; 'non-reactive factors', 'legislative factors', Nicholson, 1954; 'density-independent processes', Clark *et al.*, 1967).

In cells C and D are placed all occurrences of extrinsic relative shortage of resource, all occurrences of facultative predators, nearly all occurrences of weather and probably all occurrences of malentities, as well as the special but important occurrence of the pathogen *Myxoma*. Extrinsic relative shortages of food and other resources are commonplace in nature; and occurrences of the other events in this list are not negligible. The case-histories of *Thrips* (sections 8.2, 3.11), *Glossina* (section 3.11), *Levuana* (section 5.11), *Metriocnemus* (section 8.3) and *Oryctolagus* (section 5.21) are evidence of the importance of these 'density-independent processes'.

PART II

Practical course

10 *Methods for estimating density, patterns of distribution and dispersal in populations of animals*

10.0 INTRODUCTION

Ecologists commonly speak of the *density* of a population when they wish to refer to the number of animals in a particular area. Broadly, there are two ways of measuring the density of a population. If it is possible to count or estimate all the animals in the area it is proper to speak of the result as the *absolute density* of the population. But there are occasions when it is sufficient to know in what ratio one population exceeds another without knowing what is the actual size of either population. This is especially true when one is studying fluctuations with time in a population living in the same place. Sometimes technical difficulties may make it impossible to measure the absolute density of the population no matter how desirable this may be, and to estimate *relative densities* may be the best that one can do.

10.1 THE MEASUREMENT OF RELATIVE DENSITY

The characteristic feature of all methods for measuring *relative density* is that they depend on the collection of samples that represent a constant but unknown proportion of the total population. The sample may be collected by a conventional trap, such as a fly-trap or a mouse-trap, or by any other device that will serve the purpose. For example, when Browning (1959) needed to measure the fluctuations in a population of mealybugs that were living on the leaves of orange trees in an orchard, he collected samples of leaves and counted the mealybugs on them. It was impracticable to take the leaves strictly at random but Browning worked out a method that allowed him to take samples of several hundreds of leaves approximately at random except for the restriction that all the leaves were taken from the lower half of the tree. This method depended on the assumptions that (*a*) the leaves were picked at random, (*b*) the total number of leaves on the tree did not vary, (*c*) there was no change in the relative proportions of the mealybugs on the lower and upper parts of the tree. Provided that these three conditions are fulfilled the mean number of mealybugs on a leaf represent a constant proportion of the population on the tree. The actual density of the population could have been estimated if the number of leaves on the tree had been known. But it was impracticable to count them. It was also unnecessary because in this study relative densities were sufficient.

The *suction trap* used by Johnson (1950a) to catch aphids and other small insects floating in the air consisted essentially of a large vertical funnel, a fan driven by an electric motor, and a jar in which the insects were collected. The fan was designed to suck a constant volume of air through the funnel, about 19,000 cubic feet an hour. Such a trap, running for a known time, would collect all the aphids from a known volume of air. With sufficient number of traps appropriately distributed in the air it becomes merely a matter of simple arithmetic to estimate the actual number of aphids in the air over a particular area at a particular time. The rest of the population are on plants. If the number in the air represented a known proportion of the ones on the plant then the calculation could be extended to give an estimate of the absolute density of the population. If, on the other hand, the numbers in the air represented a constant but unknown proportion of the total population the method could provide estimates of relative density. Johnson (1952a, b, 1955) has used the records more for the study of dispersal than of density. Formerly it had been thought that the proportion of the population in the air depended on the immediate or very recent weather stimulating the aphids to fly. Although this still seems to be important Johnson has shown that the numbers in the air depend much more than had been thought on the actual density of the population on the plants.

Estimating changes in the numbers of aphids on a crop of, say, potatoes or beans may be a matter of considerable economic importance but it presents a tricky problem because the numbers may be large and they may be distributed unevenly. Banks (1954) used the following method in a crop of beans. The beans were planted in rows. A number of stems were chosen at random in each row and 'rated' into five categories merely by looking at them and judging the extent of the infestation. The five categories were zero, very few, few, many and very many.

Ten stems from each class were cut off at the base and the aphids on them were counted. From the average number of aphids on a stem in each category and from the frequency of stems of each category in the field a weighted mean number of aphids per stem was estimated. In a series of nine weekly estimates made in this way the mean number of aphids per stem increased from 20 at the beginning to 14,000 towards the end of the period. The number of stems per acre was not known, so these figures cannot be converted into an estimate of absolute density.

The light-trap that Williams (1940) used consisted essentially of an incandescent electric globe, a funnel and a jar. The light attracted night-flying Lepidoptera and other insects; the funnel directed them into the jar where they were killed. Williams ran the trap for a number of years at Rothamsted. The numbers of insects in the trap varied greatly from night to night. A substantial part of this variation was related to fluctuations in certain components of the weather which probably influenced the moths' activity in seeking out the trap. This result illustrates the general rule that the numbers caught in any trap depend partly on the density of the population that is being sampled and partly on the activity of the animals in seeking out the trap. It is desirable to find some way of distinguishing between these two components before using the results as an estimate of relative density.

Davidson and Andrewartha (1948a, b) used the statistical method of partial regression to measure the influence of temperature on the activity of *Thrips imaginis* in seeking out traps (see experiment 12.23). This is a small winged insect about two mm long. The thrips were strongly attracted to roses and these were used as 'traps' (section 3.11). The thrips were breeding in a variety of other flowers in and around the garden. But they are characteristically restless and rarely stay in one place for long. A proportion of those flying in the garden during the day settled in the roses and some of them were found there when we collected our daily sample in the morning. These records were kept for the months August–December (southern spring) for 14 years. A characteristic series of records is shown in Fig. 3.02 which represents the daily numbers of thrips per rose during 1932 (Evans, 1933).

The seasonal trend and the enormous daily fluctuations about the trend shown for this year are characteristic of all the years for which records were kept. In order to analyse the data we first transformed the numbers to logarithms; this was done chiefly because proportional changes in numbers were more instructive than additive changes. It was then possible to account for the systematic trend of numbers with time by fitting empirically a curvilinear regression of the form:

$$Y = a + bx + cx^2 + dx^3 \qquad (10.1)$$

where Y represents the logarithm of the number of thrips and x repre-

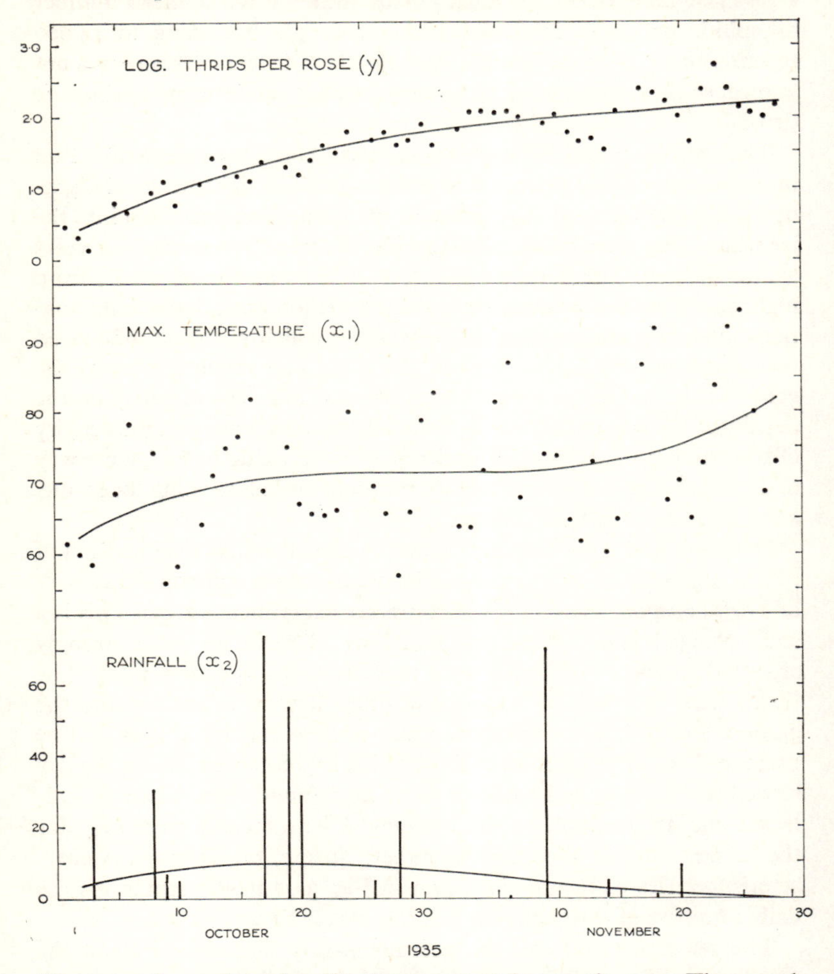

Fig. 10.01. Some of the data (for the year 1935) for equation 10.2. The smooth curves have in each instance been obtained by calculating the appropriate polynomial of the form indicated in equation 10.1. Temperature is measured in degrees Fahrenheit and rainfall in units of 0·01 in. For further explanation see text. (After Davidson and Andrewartha, 1948b.)

sents days. Figure 10.01 for 1935 is typical. The systematic trend that this curve describes is due chiefly to the natural increase of the population, but it may also be partly caused by trends in the weather. The important point is that the curve has accounted for the trend with time, whatever may be its cause, so that the residual variance (represented by the departures of the observed points from the trendline in Fig. 10.01) is independent of time. These departures are largely due to daily variations in the activity of the thrips moving into and out of the 'traps'. On a warm day more thrips moved out of the flowers in the surrounding fields, where they were breeding, and found their way into the newly opened roses, which were sampled daily.

The method of partial regression provides a way to measure the degree of association between the 'activity' of the thrips (as represented by deviations from the smooth curve) and whatever components of environment that may seem important. The final step in the analysis is to solve the equation for partial regression which may be written:

$$Y = b_1 x_1 + b_2 x_2 + b_3 x_3 \qquad (10.2)$$

where Y represents the departure of the observed data from the smooth curve in Fig. 10.01, $x_1, x_2, x_3, \ldots$ represent temperature, humidity, and other components of environment also expressed as departures from the smooth curves (Fig. 10.01); and $b_1, b_2, b_3, \ldots$ are the coefficients for partial regression such that b_1 measures the association of Y with x_1, independent of any relations that might exist between these two and any other variables in the equation.

In the present example it was found that the independent influence of temperature was highly significant and that, on the average, the numbers of thrips in the flowers on any one day increased by 5% for every increase of 1°C in the daily maximal atmospheric temperature.

The analysis also showed that the influence of temperature was the same whether there were few or many thrips in the population. Apparently the movement of the thrips to and from the breeding places to the roses was independent of the density of the population. It was not the outcome of jostling or any other manifestation of numbers; it was rather likely to have been the outcome of an innate tendency to disperse possessed equally by thrips living in solitude and those living in a crowd. This tendency may be influenced by several components of weather of which temperature is perhaps the most important (section 10.4).

10.2 THE MEASUREMENT OF ABSOLUTE DENSITY

10.21 *Counting the whole population*

The most direct way to find out how many animals are living in an area is to count them all. Errington (1945) kept a complete record during 15 winters of the number of bobwhite quail, *Colinus virginianus*, on 4,500 acres of farmland in Wisconsin by counting either the birds or their 'sign'. The sign left by the birds in the snow supplied a lot of information apart from their mere numbers. Similarly with the muskrat, *Ondatra zibethicus*, Errington (1943, 1946) was able to count all those living in certain marshes in Iowa. Again the counting was done largely by reading sign.

Kluijver (1951) reported a 40-year study of the ecology of the great tit, *Parus major*, in 130 acres of mixed woodland in Holland. The tits normally use holes in trees for nesting during summer and roosting during winter. But they will readily use artificial nesting boxes instead. This behaviour allowed Kluijver to inspect the birds at night, to mark them with leg-bands and to count them all.

Usually there is a better chance of counting the whole population when dealing with vertebrates, especially birds and rodents, than with insects and other invertebrates. This method has many advantages but the occasions on which it can be used are relatively infrequent even with vertebrates. Usually one has to rely on some method of sampling in which only a small proportion of the population is seen. Broadly there are two methods available. Either one counts every animal in a quadrat that represents a known *proportion* of the total population. Or one captures, marks and releases a known *number* of animals. The total population is estimated from the proportion of marked ones that are subsequently recaptured. Of course, the details vary widely according to the behaviour of the particular animals that are being studied. I shall discuss these two methods in sections 10.22 and 10.23.

10.22 *The use of quadrats*

In the simplest form of this method the quadrats are merely small areas chosen at random from a larger area which contains the whole population. The conditions that must be fulfilled are that the population of the quadrat must be known exactly and the area of the quadrat relative to the whole area must be known. Details of the devices that have been used for different species vary greatly. I shall mention only a few.

Salt *et al.* (1948) estimated the number of arthropods in an acre of soil by cutting out cylinders of soil four inches in dia. and 12 inches deep. They invented a special technique for pulverizing the soil without destroying the animals in it. The soil was then washed through a series of sieves and the fractions containing plant debris were further sorted by a flotation process using an interface between paraffin and an aqueous solution of magnesium sulphate. On one occasion Salt *et al.* found that a sample of 20 quadrats (i.e. 20 cylinders of soil four inches in dia.) contained on the average 2,138 animals per quadrat. Multiplying this figure by 498,960 (the number of quadrats that could be taken from one acre) gives an estimated population of 1066·6 million arthropods per acre. Of these 666 million were mites and 379 million were insects.

Often it is possible to make use of some peculiarity in the behaviour of the animal in order to collect them from or count them in the quadrats. Madge (1954) used quadrats one foot square to estimate the density of the population of *Oncopera fasciculata*. The caterpillars live in vertical burrows in the soil under pastures of clover and grass. The mouth of the burrow is difficult to find because it is well camouflaged. Even clearing away the herbage does not help because this blocks the mouth of the burrow and makes it invisible. But if the herbage is cleared away and then a flat board is pressed down on the soil the caterpillars will clean out their burrows and spin a silken cap to them. When the board is lifted next day the mouths of the burrows stand clearly revealed and can be counted easily.

The small fly, *Metriocnemus hirticollis* (section 8.3), breeds in sewage filter-beds near Leeds. The eggs, larvae and pupae are found in the filter-beds but the adults emerge into the air to mate. Mating does not occur casually on or near the surface of the filter-bed but only after an elaborate nuptial flight that involves the formation of a nuptial swarm. The flies always seek to escape from the surface of the filter-bed. Consequently Lloyd (1941) was able to trap them by placing a funnel-like frame one foot square on the surface of the filter-bed which led the flies upward into a jar in which they were trapped. Multiplying the number of flies caught in a trap by the surface area of the filter-bed in square feet would give an estimate of the absolute density of the population.

The sheep tick, *Ixodes ricinus*, takes about three years to complete its life-cycle. Most of this time is spent near the surface of the ground, usually concealed in a dense mat of vegetation and decaying organic matter where it is difficult to find. But during the spring each year there is a period of several weeks during which the ticks come up for perhaps a day or two at a time to sit on the tips of grass-stems and other exposed places. If a sheep brushes it at this time the tick will cling to the sheep. Milne (1943) found that the ticks would also cling to a woollen blanket if it were dragged over the grass. Of course not all the ticks were exposed at once but by 'dragging' the same area repeatedly Milne considered that he could collect all or nearly all the ticks. He chose a number of quadrats of known area and 'dragged' them thoroughly. In this way he estimated the total population of ticks on a particular farm at between 140,000 and 230,000.

The reliability of the estimates made from counting the numbers in quadrats depends on how representative the quadrats are of the whole population and on how close one gets to seeing all the animals in the quadrat. The latter question is a matter of technique and must be resolved on its merits for each particular species. Salt *et al.* pointed out that their method was likely to miss any animal that could pass through a 0·1-mm sieve; they thought that the number missed in this way might be quite substantial. Milne could count only the ticks that had emerged from the 'mat' and were picked up by his blanket. He 'dragged' the same area repeatedly during the active season and probably left very few behind. Madge could count only the *Oncopera* that repaired their tunnels when he lifted the board. But he could check his results by digging out the soil and sieving it. It seemed that he would miss very few provided that he sampled when the larvae were not likely to be moulting.

How well the sample of quadrats may represent the population is largely a statistical matter. The size of the sample (the number of animals counted) that is required for a particular level of accuracy depends on the variance. Larger samples will be needed for the same degree of accuracy when the variance is large. The variance will be large when the animals are distributed patchily (section 10.4). For certain sorts of 'contagion' the size of the sample required for the same level of accuracy increases roughly in proportion to the size of the 'cluster'. If the 'patchiness' of the population follows a pattern that can be recognized it may be practicable to take advantage of this knowledge to improve the accuracy of the estimate by the method of stratified sampling (section 10.51).

10.23 *The method of capture, marking, release and recapture*

If, from an area that has been selected for study, M individuals have been captured, marked and then released again, and if on a second occasion N individuals are captured, including R marked ones, then the population in the area may be estimated as:

$$P = N \times \frac{M}{R} \tag{10.3}$$

The idea expressed by this equation is simple enough in principle but it implies two assumptions that need to be discussed.

(1) Equation 10.3 implies that the marked ones after they were released distributed themselves homogeneously with respect to the unmarked ones which had not been caught. The equation also implies that on subsequent occasions any marked one has the same chance of being recaptured as any unmarked one has of being captured.

(2) Equation 10.3 also implies that the recapturing is done immediately after the releasing, or at least before there has been time for any marked one to die or leave the area, or for any immigrant to enter the area. In practice this cannot be done because the marked ones must be allowed time to mix with the unmarked ones.

There are also two technical difficulties that should be mentioned.

(1) When traps are used, especially in an area that is not limited by sharp ecological boundaries, difficulties arise with respect to the placing of traps relative to the boundaries of the area.

(2) It is difficult to attribute a precise variance to the estimates that are given by this method because the calculations are tedious and difficult.

Dowdeswell *et al.* (1940) estimated the density of a population of the butterfly *Polyommatus*. The butterflies were caught with a net, marked and released in the same place that they were caught. Jackson's (1939) method of capturing and releasing the tse-tse fly, *Glossina morsitans*, was similar. There would seem to be little risk of disturbing the distribution of the marked relative to the unmarked when this method is used. But often it may be necessary to use some sort of trap. If the trap lures the animals from any distance it results in an artificial aggregation of a part of the population. These individuals are marked and the problem is to get them back into the population so that they are distributed in the same way as the unmarked ones. It is not sufficient merely to scatter them at random to the points of the compass or to any other geometrical co-ordinates because the unmarked part of the population may not be distributed in this way. All that can be done is to release them and hope that, in settling down, they will conform to the same pattern of distribution as the unmarked ones.

Evans (1949) exposed a number of traps in a building where there was a population of *Mus musculus*. Certain individuals were captured many times whereas others were not caught once. The differences were too great to be attributed to chance. It was clear that certain mice were much more likely to be caught than others. This placed an intolerable bias on the estimates of the density of the population and the exercise had to be abandoned. Even when more direct methods are used for capturing the sample there may be a risk of bias from this cause. For example, if the more sluggish animals are more easily caught than the more lively ones, then the more sluggish would also be more easily recaptured and this would tend to place a bias on the estimates of density. Even if it were true only that the older ones were more easily caught than the young ones, this would introduce bias because the marked ones must be older than the unmarked. There is no standard way to get over this difficulty; each case must be treated on its merits even to abandoning the exercise if the problem cannot be solved.

If the marked redistributed themselves quite homogeneously with respect to the unmarked the recapturing could be done systematically or haphazardly over the whole area or part of it without any risk of introducing bias to the results. But in so far as the marked are not distributed homogeneously with respect to the unmarked (and in practice they rarely will be) the only way to reduce the risk of bias is to randomize the positions of the recaptures over the whole area.

Equation 10.3 implies that the recapturing is done immediately after releasing but in practice some delay must be allowed while the marked animals are given time to disperse and settle down. Consequently it is necessary to extrapolate backwards from the date when they are recaptured to the date when they were released. This makes it necessary to recapture on a number of occasions instead of only once, or alternatively to mark on a number of occasions and recapture once. The former was called the 'positive' and the latter the 'negative' method by Jackson (1939); both lead to the same result within the limits of sampling errors. The calculations are best explained by reference to a particular example and I shall use some figures that were published by Jackson (1939) for a population of tse-tse fly, *Glossina morsitans*.

The figures are set out in Table 10.01. First consider the lower part of the table covering the period from t_0 to t_6. These are the data required for the 'positive' method of estimating the number in the population on date t_0. The table shows that on date t_0 1,558 flies were captured, marked and released. On date t_1 a total of 1,547 flies were caught, including 101 that had been marked and released on date t_0. On date t_2

Table 10.01. *Jackson's (1939) data for Glossina morsitans (slightly adjusted)**

Date of capture or recapture	Marked and released	Total captured	Recaptured	Number of flies corrected recaptures y	Estimated population	
					Negative	Positive
t_{-6}	1,262		1	0·0051		
t_{-5}	1,086		5	0·0296		
t_{-4}	1,299		17	0·0840		
t_{-3}	1,401		28	0·1283		
t_{-2}	1,198		48	0·2569		
t_{-1}	1,183		70	0·3798		
t_0	1,558	1,558			12,800	13,200
t_1		1,547	101	0·4190		
t_2		1,307	81	0·3978		
t_3		603	24	0·2555		
t_4		1,081	31	0·1841		
t_5		1,261	10	0·0509		
t_6		1,798	11	0·0393		

* Jackson's original figures show 1,582 marked on date t_0 for the positive method and 1558 caught on date t_0 for the negative method. The discrepancy is due to 24 young unfed flies which, according to the conditions that Jackson had made, were eligible for marking but not for counting in the 'negative' catch. In order to simplify the arithmetic I have ignored these 24 young flies and adjusted the recaptures for the positive method as if 1,558 had been marked and released on day t_0.

the catch totalled 1,307, including 81 of the marked ones. The numbers in these two columns vary from day to day. In order to simplify the calculations we correct the figures for recaptures to what they would have been if 100 had been marked on date t_0 and 100 captured on each successive date. We call these corrected figures $y_1, y_2, \ldots y_6$. They are calculated as follows:

$$y_1 = \frac{101}{1} \times \frac{100}{1{,}558} \times \frac{100}{1{,}547} = 0{\cdot}4190 \qquad (10.4)$$

We require to find by extrapolation from this series y_0 which we think of as the (corrected) number that would have been recaptured on t_0 if we had been able to recapture immediately after releasing. In order to recognize the difference between this theoretical estimate and the empirical series of y we call our estimate a_0 instead of y_0. In order to estimate a_0 we first require to calculate r_+ which is the weighted ratio of $\frac{y_n}{y_{n-1}}$. When there are no more than 3 or 4 values for y (i.e. when recaptures were made on no more than 3 or 4 dates) it may be necessary to estimate r_+ from equation 10.5:

$$r_+ = \frac{y_2 + y_3 + \ldots + y_n}{y_1 + y_2 + \ldots + y_{n-1}} \qquad (10.5)$$

But when there are sufficient values of y it is better to use equation 10.6 which carries the smoothing further:

$$r_+ = \sqrt{\left(\frac{y_3 + y_4 + y_5 + \ldots + y_n}{y_1 + y_2 + y_3 + \ldots + y_{n-2}}\right)} \qquad (10.6)$$

In the present example the value for r_+ by equation 10.6 is:

$$r_+ = \sqrt{\left(\frac{0{\cdot}2555 + 0{\cdot}1841 + 0{\cdot}0509 + 0{\cdot}0393}{0{\cdot}4190 + 0{\cdot}3978 + 0{\cdot}2555 + 0{\cdot}1841}\right)}$$
$$= 0{\cdot}6494$$

The value for a_0 is given by equation 10.7:

$$a_0 = \frac{y_1 + y_2 + \ldots + y_{n-1}}{r_+} - (y_1 + y_2 + \ldots + y_{n-2}) \qquad (10.7)$$

Because $\frac{y_1}{r_+}, \frac{y_2}{r_+}$, etc. may be regarded as estimates of y_0, y_1, etc., taking into account the weighted estimate of r_+, it is convenient to regard equation 10.7 as being equivalent to:

$$y_0 + (y_1 + y_2 + \ldots + y_{n-2}) - (y_1 + y_2 + \ldots + y_{n-2}) \qquad (10.8)$$

The two terms in brackets cancel out. For the present example:

$$a_0 = \frac{0{\cdot}4190 + 0{\cdot}3978 + 0{\cdot}2555 + 0{\cdot}1841 + 0{\cdot}0509}{0{\cdot}6494} - (0{\cdot}4190 + 0{\cdot}3978 + 0{\cdot}2555 + 0{\cdot}1841)$$
$$= 0{\cdot}757$$

Now all the values of y, and therefore of a_0 have been corrected as if 100 flies were marked on date t_0 and 100 were captured on all subsequent dates. Therefore, substituting these values in equation 10.3, we have:

$$P\text{ (on date } t_0) = \frac{100}{1} \times \frac{100}{a_0} \qquad (10.9)$$

For the present example:

$$P_0 = \frac{100}{1} \times \frac{100}{0{\cdot}757}$$
$$= 13{,}200$$

This is the estimate, by the positive method, of the population on date t_0.

Now consider the top half of Table 10.01 which sets out the data for the negative method. In this method we consider recaptures on date t_0 of flies that were marked and released on previous dates t_{-1}, t_{-2}, . . ., t_{-n}. Thus in column 4 the number 28 opposite t_{-3} indicates that of the 1,401 flies that were marked and released on date t_{-3}, 28 were recaptured on date t_0. Similarly 17 out of the 1,299 that were marked and released on t_{-4} were recpatured on t_0. The total catch on date t_0 including all those that had been marked on the six previous occasions was 1,558.

The equations for calculating $y_1, y_2, \ldots, y_n$ are the same as in the positive method. For example:

$$y_{-4} = \frac{17}{1} \times \frac{100}{1{,}299} \times \frac{100}{1{,}558}$$
$$= 0{\cdot}0840$$

And r_- and a_0 are calculated from the same equations as in the positive method. For the present example:

$$r_- = 0{\cdot}5394$$
$$a_0 = 0{\cdot}780$$
$$P_0 = 12{,}800$$

In this method we think of a_0 as the (corrected) number that would have been recaptured on t_0 if they had been marked and released on that day and recaptured immediately, before any of the marked ones had been lost by death or emigration. The two estimates of P_0 agree quite well.

Boguslavsky (1956) described a method of capture, marking, release and recapture that is especially suitable for small populations. One individual is captured, marked distinctively and released. Subsequently samples of one are drawn repeatedly from the population and released again. On the first occasion that any individual is seen it is marked distinctively before being released. Suppose that in one area, after repeating this procedure a number of times, only three animals have been seen and marked. If the procedure is continued until 13 recaptures have failed to reveal any other animal than these three then the odds against there being more than three animals in the area are nine to one, always provided that all the animals in the area are equally likely to be captured. Similarly if 15 animals were seen and marked it would be necessary to repeat the sampling for 78 consecutive times without seeing a stranger in order to establish odds of nine to one against there being more than 15 animals in the area. Although the arithmetic becomes rather cumbersome for larger numbers Boguslavsky showed how to calculate the number of negative trials that are necessary in order to establish any size of population at any level of probability. Subject to the restrictions mentioned above, this method seems promising for small local populations especially in wildlife studies. Still other methods have been developed for estimating the number of fish in a lake or other enclosed area. For an example of how these methods may be used see Gerking (1952).

More advanced methods than the rather elementary ones that I have described in this section have been developed for analysing the results of capture, marking, release and recapture. They give better estimates of birth-rates, death-rates and densities; also expressions for the variance of r_{+}, r_{-} and a have been worked out which allow the precision of estimates to be known. These methods may be found in papers by Jackson (1940, 1944, 1948), Bailey (1951, 1952), Leslie (1952) and Leslie and Chitty (1951).

10.3 THE MEASUREMENT OF 'AGGREGATION' IN NATURAL POPULATIONS

In his essay on 'population interspersion' Elton (1949) pointed out that no habitat is homogeneous. Particular sorts of animals may be restricted to particular 'minor habitats' scattered through the larger area. Thus in a mixed deciduous wood there may be a particular sort of insect that can live only on oak trees. This immediately imposes a 'patchiness' on the distribution* of the animals: they are to be found only where there is an oak tree. On close examination, any habitat, even such a seemingly uniform one as a grove of citrus, will reveal more or less heterogeneity with respect to the sorts of places where animals may live. The trees may differ with respect to genetical constitution, health, age, etc.; even those that are otherwise uniform must have a north and a south face. This heterogeneity in habitats is likely to be reflected in 'patchiness' in the way that the animals are distributed in the area.

But if one narrows the study-area down to a small and uniform area such as Elton called a 'micro-habitat' – the leaves on the western face

* In this section 'distribution' is used with its statistical meaning. The relative frequency of quadrats containing 0, 1, 2, . . . animals forms a 'frequency distribution'. The context is quite different from that in Chapter 1 where 'distribution' was used in its ecological meaning to connote 'area'.

of a tree, or six square inches under its bark – one will still find, more often than not, that the animals are distributed according to some pattern that is different from what would be expected if they had been scattered at random over the area. Often the explanation of the pattern lies in the behaviour of the animals or in some events in the history of the population.

I shall discuss several methods for measuring these 'patterns of distribution' in section 10.31.

10.31 *A test for randomness: the Poisson series*

Imagine a uniform area subdivided into several hundred quadrats. Imagine that animals have been scattered over the area at random to each other and to the area. This means that each animal, as it was released, was just as likely to settle in any one quadrat as in any other and that its chance of settling in a quadrat was independent of whether this quadrat was already occupied by one or more animals or not. The frequency distribution setting out the number of quadrats with 0, 1, 2, 3, . . . animals on them is likely to conform, within the limits of random sampling, to the expansion of the binomial $(p + q)^n$ where p is the chance that an animal will settle in any quadrat; q is the chance that it will not; because the animal must either settle or not settle in that quadrat $p + q = 1$; n is the total number of animals. Provided that n is large, that p is small relative to q and that p is of the same order of magnitude as $\frac{1}{n}$ the expansion of the binomial $(p + q)^n$ approximates to the Poisson series which is given by:

$$P_x = e^{-m}\left(1, m, \frac{m^2}{2!}, + \ldots \frac{m^x}{x!}\right)$$

In this expression P_x is the probability of getting 0, 1, 2, . . . animals in a quadrat. The frequency with which these numbers would be expected in an experiment in which N quadrats had been counted is given by multiplying the probabilities by N. For values of m greater than 10 the Poisson rapidly approaches the normal distribution.

It is characteristic of the Poisson series that the variance equals the mean so that the ratio $\frac{S(x - \bar{x})^2}{(n - 1)\bar{x}} = 1$; and for a Poisson series the expression $\frac{S(x - \bar{x})^2}{\bar{x}}$ estimates χ^2 with $n - 1$ degrees of freedom. Values of less than unity for the first expression indicate that the observed distribution from which it was calculated is more uniform, and values greater than unity indicate that the distribution is more variable than the Poisson series. Departures from unity in either direction increase the significance of χ^2. Methods for calculating the Poisson series and the tests for agreement with the observed series, as well as the meaning to be attached to the tests for significance, are discussed in section 10.5.

Table 10.02 shows how this method was used to study the distribution of a marine worm of the family Eunicidae. The worms live in burrows

Table 10.02. *The distribution generated by counting the numbers of worms (Eunicidae) in quadrats one foot square compared with the Poisson series*

Worms per quadrat (x)	Number of quadrats (f)	Poisson series	$\frac{(O-E)^2}{E}$
0	309	253·96	11·93
1	241	291·42	8·72
2	138	167·21	5·10
3	68	63·95	0·26
4	30	18·34	7·41
5	10	4·21	
6	3	0·81	
7	0		16·10
8	0	0·16	
9	1		
Total	800	800	49·52

$$\frac{s^2}{\bar{x}} = \frac{1·590}{1·148} = 1·39$$

$$P(\chi_4^2) < 0·001$$

in the mud just above the level of low tide. They were counted in quadrats one foot square. The large value for χ^2 ($P < 0·001$) indicates a significant departure from the Poisson series; and the value of 1·39 for the ratio of variance to mean indicates that the observed figures were more variable than the Poisson series; there were too many quadrats with 0 and more than three worms and not enough quadrats with one or two worms in them. Such a pronounced 'clumping' is unlikely ($P < 0·001$) to have occurred from a random scattering of the worms. Some other hypothesis is needed.

10.4 DISPERSAL

Broadly speaking, animals have three ways of dispersing – by drifting with currents of air or water, by swimming, walking or flying, and by clinging to some moving object; the moving object may be some sort of animal, as a tick may cling to a sheep or it may be some article of commerce, as a grain weevil is dispersed in a sack of grain.

10.41 *Dispersal by drifting*

Drifting cannot always be clearly distinguished from swimming or flying because many species may get a start this way or even maintain themselves afloat or aloft by their own efforts while being carried by the current. But it is worth making the distinction because this is the chief method of dispersal for so many small arthropods that are serious pests, and for most of the small animals that live in the sea; also many of the larger animals in the sea have planktonic larvae that disperse in this way.

10.411 Dispersal of the bean aphis *Aphis fabae*

The bean aphis lives by sucking sap from the bean, *Vicia faba*, especially from young plants. It is a serious pest because it sometimes becomes very abundant and retards the growth of the crop. Aphids as a group do a lot of damage to horticultural crops not only by sucking sap but also by spreading virus diseases. Therefore a knowledge of the dispersal of aphids is relevant to many important practical problems in horticulture. The principles that apply to the dispersal of *A. fabae* are likely to apply to the dispersal not only of other sorts of aphids but also to many other small animals that drift in the air. Johnson and his colleagues at Rothamsted have made a thorough study of the dispersal of *A. fabae* (Johnson, 1950b; 1952a, b; 1954; 1955; 1956; 1957; Johnson and Taylor, 1957; Taylor, 1957).

These classical papers are important reading not only for the study of dispersal but also because they are an elegant demonstration of the slaying of a beautiful (and well entrenched) hypothesis by empirical fact.

10.412 Dispersal of wingless insects, mites and spiders

The principles that have been established for the dispersal of *A. fabae* are likely to be helpful with many other species that use this method of dispersal. Coad (1931) dragged a net behind an aeroplane and estimated that the air between 50 feet and 14,000 feet above one square mile near Tallulah, Louisiana, contained 25 million insects and other small animals. Most of them were in the first 1,000 feet but some were found at all levels up to 14,000. The larger, heavier ones and the stronger fliers were closer to the ground; the smaller ones, the weaker fliers and the wingless ones constituted most of the 'catch' at the higher altitudes.

There is convincing evidence that the scale insect *Eriococcus orariensis* which recently spread dramatically through the indigenous stands of *Leptospermum scoparium* in New Zealand was originally blown from Australia, a distance of more than 1,000 miles across ocean. This species occurs on *Melaleuca* in Australia; it is most unlikely to have been carried accidentally to New Zealand either by tourists or by traders. Moreover the original outbreak was traced back to a lonely and inaccessible mountain valley where it was most unlikely to have been carried by any means other than wind. The speed with which it subsequently spread in New Zealand is also good evidence of the effectiveness of its dispersal by wind (Hoy, 1954).

Duffey (1956) studied the dispersal of nine species of linophiid spiders that are common in the Wytham estate near Oxford. They all have one generation a year. Some species disperse in the spring, others in the autumn. But with all of them, dispersal by drifting with the wind is a normal part of their behaviour which recurs at the same stage of the life-cycle in every generation.

I have been impressed by the ubiquity of the small sap-sucking mites *Tetranychus* on the plants in my garden. The mites breed on a variety of herbaceous garden plants and weeds. It is most striking that they usually manage to colonize a seedling within a few days of its appearance above ground. It seems that they are, through most of the summer, floating in the air in such large numbers as to make the early discovery of any new plant a near certainty.

10.413 The dispersal of locusts

Locusts can fly powerfully as any one who has tried to catch them with a net may testify. Yet, strangely enough, most of the large-scale dispersal of locusts is done by drifting with winds at high altitudes. When a swarm of locusts has been resting for the night they remain on the ground next morning, more or less quiescent, until the temperature rises above a certain threshold which may vary with the turbulence of the air. Usually they fly more readily when the air is more turbulent. So they are often carried upward by ascending currents of air until they are several thousands of feet above the ground (Rainey and Waloff, 1951). A locust that is flying close to the ground usually seeks shelter if the wind becomes strong. But a locust that is high enough to be out of sight of the ground no longer perceives changes in the direction or the velocity of the wind and it may remain aloft for a long time and be blown hundreds of miles.

Waloff (1946a) traced the path of one swarm of *Schistocerca gregaria* from Morocco to Portugal. Waloff (1946b) analysed many records of the movements of swarms of *S. gregaria* in eastern Africa; and Rainey and Waloff (1948) did the same for the area around the Gulf of Aden. Most of the long migrations were the outcome of the locusts having been blown with the prevailing winds, especially those at altitudes of 2,000 to 3,000 feet. There was no evidence that the migrations led, except by chance, to favourable places for breeding; but when, in the course of their wanderings with the wind, the locusts arrived at a place that was moist enough, they would develop to sexual maturity and lay eggs.

10.414 The dispersal of strong-flying insects by winds near the ground

Coad (1931), see above, found that most of the large strong-flying insects in his 'catches' were caught at lower levels and it seems that the dispersal of many insects of this sort is regularly helped by surface winds. Barber (1939) found the fly *Euxesta stigmatias* breeding in an isolated field of sweet corn in Florida in circumstances that strongly indicated that they had been blown in from the West Indies. Dorst and Davis (1937) traced the dispersal of the beet leafhopper *Eutettix tenellus* from winter breeding grounds across 200 miles of desert (where there was no food for the leafhoppers except on the experimental plots used for trapping them) to the beet-growing areas of Utah. Felt (1925) pointed out that the moth *Heliothis armigera* rarely, if ever, overwinters north of southern Pennsylvania and the Ohio river, but it sometimes becomes abundant on sweet corn in southern Canada. Felt also listed a number of other species from the tropics or the sub-tropics which were unable to survive a northern winter, yet from time to time were found breeding in Canada during summer. It is likely, in all these instances, that the animals had been blown north by winds but we do not know whether they travelled in the upper air; it seems more likely that they may have been blown by winds near the ground.

10.415 The dispersal of marine species that drift with currents

Most of the smaller animals in the sea disperse by drifting. Many species have planktonic larvae that are especially adapted for drifting with the current. Some members of the permanent plankton are so characteristic of particular currents that they have been called 'current indicators' (Russell, 1939).

Certain barnacles and oysters that live in tidal rivers and estuaries have planktonic larvae whose behaviour shows special adaptations which allow the population to persist in these places despite the fact that the net flow of water is outwards towards the ocean. Like the aphids, they make use of turbulence. The denser water near the bottom tends to have a net flow inland. Barnacles, as they approach the settling stage, tend to swim down and congregate near the bottom (Bousfield in Clarke, 1954). The larvae of oysters, as they get a little older, tend to cling to the bottom during the ebb tide and rise into the water during the flood tide (Carricker, 1951).

The dispersal of the European eel involves both drifting and swimming at different stages in the life-cycle (Schmidt, 1925). The spawning of the eel has not been described. The small flat planktonic larvae can first be found floating in the water in an area south-east of Bermuda. They drift with the Gulf Stream. After about three years they reach the coast of Europe or North Africa. They metamorphose to elvers which swim up the rivers. After several years' development in fresh water the mature eels leave the rivers and return to the same area that they came from as larvae. It is not known how they find their way back to the same place. It seems likely that they do it by swimming rather than drifting.

10.42 *Dispersal by swimming, walking or flying*

Certain butterflies, fish, birds and whales are known to 'migrate' great distances – hundreds or even thousands of miles. The animals may undertake these long journeys quite regularly as a characteristic feature of their lives. They may resemble the eel (section 10.415) in that the whole population, or the whole breeding population, moves from one extreme of the distribution to the other. Many birds, like the eels, breed only at one end of the distribution but, unlike the eels, the same individuals may make the journey repeatedly twice every year (Mathews, 1955; Hochbaum, 1955). The butterfly *Danaus plexippus* breeds in Ontario during the summer; each autumn they migrate southwards.

They overwinter in southern North America. Beall (1946) showed that the immigrants that returned to Ontario the next summer were not the same individuals that had left in the autumn but were almost certainly their progeny. Seveal other species of butterflies in Europe behave similarly. Migration is a large and specialized subject so I shall not discuss it in this book.

10.421 The measurement of dispersal

Dobzhansky and Wright (1943, 1947) described a method for measuring the rate of dispersal of animals that disperse chiefly under their own power. They used *Drosophila pseudoobscura*. They arranged a series of traps in two straight lines that bisected each other at right angles

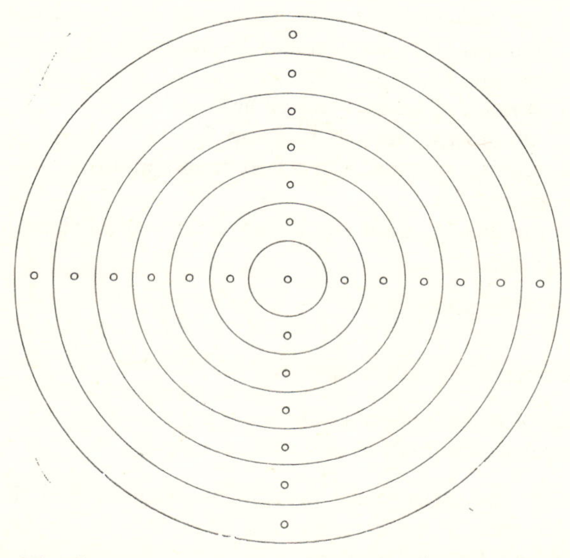

Fig. 10.02. The distribution of traps in a 'cross' as used by Dobzhansky and Wright to measure the dispersal of *Drosophila*. Theoretically each set of four traps measures the density of the population in an annulus.

(Fig. 10.02). They then set free about 4,000 *Drosophila* at the place where the two lines of traps intersected. Subsequently they counted (and released) the *Drosophila* at each trap on each of nine consecutive days. The number counted at the central trap was taken as an estimate of the population in the central circle and the numbers counted at the four traps in each annulus were taken as an estimate of the population in the annulus. The mean density in each annulus may be regarded as the frequency with which individuals occur at that distance from the centre (point of origin).

If we imagine that these frequencies have been arbitrarily divided between two directions from the centre (instead of four), the results may be set out in the conventional form of a frequency distribution whose mean has been arbitrarily set at zero (the position of the central trap, Fig. 10.03). Irrespective of whether four or some other number of traps was used to measure the density of the population in an annulus, the variance of r (the distance travelled by an animal) may be estimated as follows.

Let $r =$ the distance of any trap from the centre.
Let $d =$ the diameter of the central circle and the width of each annulus. For convenience let us work in units of d, i.e. let us call $d = 1$.

Let $\lambda = \dfrac{\text{number of animals in unit area}}{\text{number of animals in a trap}}$.

Let $f =$ number of animals in a trap in an annulus.
Let $\bar{f} =$ mean of f.
Let $c =$ number of animals in a trap in the central circle.
Let $\bar{c} =$ mean of c.

The area of central circle $= \pi\left(\dfrac{d}{2}\right)^2$.

The number of animals in this area $= \dfrac{\pi d^2 \lambda \bar{c}}{4}$.

The area of an annulus of radius $r = 2\pi r d$.
The number of animals in this area $= 2\pi r d \lambda \bar{f}$.

To obtain the variance we proceed as usual to find the sum of squared deviations from the origin and divide this by the total number of observations. In this instance:

$$\text{The variance of } r \text{ (i.e. } s^2) = \frac{S 2\pi r d \lambda \bar{f} r^2}{S 2\pi r d \lambda \bar{f} + \pi \dfrac{d^2}{4} \lambda \bar{c}} \text{ (where } d = 1)$$

$$= \frac{S\bar{f}r^3}{S\bar{f}r + \dfrac{\bar{c}}{8}}$$

The results of Dobzhansky and Wright's experiment are shown in Table 10.03. The variance for each of the nine days is shown in the third column; the variance was first calculated in trap-units and then converted to square metres.

An increase from 3,200 to 10,500 in nine days gives an average of 811 per day. But the daily increments in variance that were actually observed varied from −1,400 (an unrealistic value almost certainly due to sampling errors) to 3,200. The variability in the daily increments in variance was related to temperature. Dobzhansky and Wright found a significant correlation between T and i. The increase in s^2 was greater by 924 m² for each 1°F increase in temperature. But the relationship was not truly linear. The flies moved very little when the temperature was below 15°C. When the temperature in the evening (which is the

Table 10.03. *The dispersal of Drosophila pseudoobscura; variance in square metres, dispersal in metres**

Day	Temp. °C (T)	Variance s^2 in m²	Daily increment in s^2 (i)	Standard deviation × 0·67 in m ($s \times 0{\cdot}67$)
1	13	3,200		373
2	19	5,000	1,800	474
3	19	7,400	2,400	561
4	15	7,200	−200	569
5	13	6,100	−1,100	523
6	18	8,000	1,900	599
7	17	8,700	700	625
8	17	11,900	3,200	731
9	15	10,500	−1,400	687

* After Dobzhansky and Wright (1947).

time when the flies are most active) was about 20°C, the flies travelled on the average, about 120 m; with temperatures around 25°C they probably travelled about 200 m in a day. There is no evidence that it was greater at the beginning of the experiment than at the end. Apparently the rate of dispersal was independent of the degree of crowding. On the other hand, the presence of a few high frequencies near the tail of the distribution in Fig. 10.03 shows that the flies were heterogeneous with respect to 'dispersiveness'; i.e. the sample included a few individuals that dispersed more actively than the majority. This feature may be characteristic of many species because it has been observed in most of those that have been studied carefully.

In any normal curve half of the population is separated from the origin by a distance equal to or greater than 0·67 × the standard deviation. Thus the fifth column in Table 10.03 shows for each day the distance away from the centre that half the flies would have reached or passed. This estimate depends on the theory of the normal curve. In the present instance the observed frequencies differ rather strongly from a normal curve because of the heterogeneity that I mentioned above. So the figures in column five are likely to contain substantial

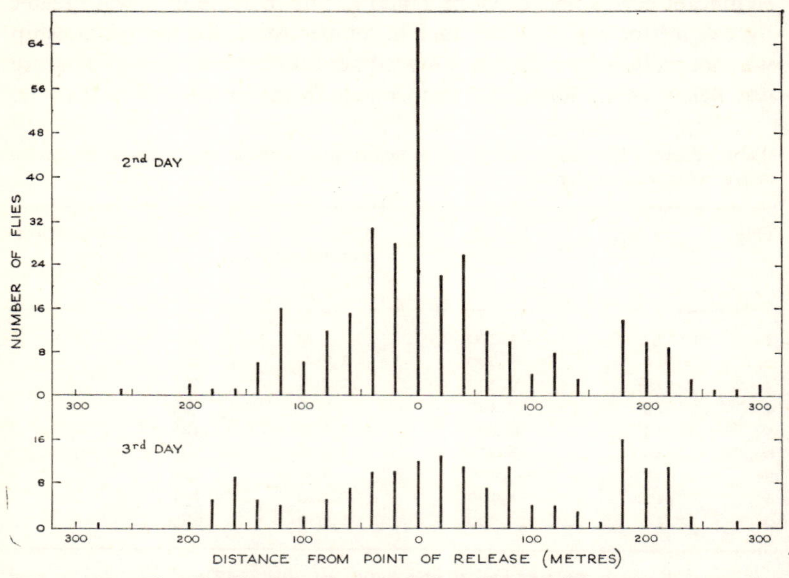

Fig. 10.03. The distribution of marked flies (*Drosophila*) on two successive days in one of Dobzhansky's experiments. See text for explanation. The unit for the abscissae is 20 m, which was the distance between traps (Fig. 10.02). The ordinates show the mean number of flies in a trap. (After Andrewartha and Birch, 1954.)

errors. So long as this limitation is remembered no harm is done in presenting them as a rough measure of the dispersal of *Drosophila pseudoobscura*.

Dobzhansky and Wright used one trap in the central circle and four in each of the annuli. This design can be improved by using several traps in the central circle and varying the number of traps in the annuli in proportion to their area. The accuracy of the estimate of density depends on the number of animals caught, not on the number of traps used. Therefore it is better to use more traps where the density is low (section 10.22). Moreover, the frequency in each annulus is multiplied by the square of the distance from the centre. So errors in the estimates for outer annuli are magnified more than those for inner annuli. Experiments of this design that can be done in class are described in section 10.7.

10.43 *Dispersal by clinging*

This is the usual way of dispersal for parasites; for some, like the sheep tick *Ixodes ricinus*, it is the only way; for others like some of the parasitic worms, it may be supplemented by a limited amount of movement, or by other means, during the free-living stages.

Modern developments in transport and trade have increased the opportunities for dispersal afforded to the wide variety of animals that may attach themselves to or conceal themselves in the things that men transport in the course of trade. Elton (1958) described a number of 'invasions' that have occurred this way.

10.44 *The special importance of dispersal in insect predators*

There is general agreement among students of biological control that a predator should be able to 'out-breed' and 'out-disperse' its prey if it is to have a reasonable chance of reducing its prey to low numbers (section 5.11). Flanders (1947) made this point referring generally to hymenopterous 'parasites' as agencies of biological control and particularly to *Metaphycus helvolus* which effectively controls the scale insect *Saissetia oleae* in some regions. Flanders first pointed out:

> In many cases it is known that the population level of a host species is determined by the host-searching capacity of the free-living gravid females of the parasitic insect. The actual level attained is determined by those attributes of the female parasite that contribute to its host-searching capacity.

He then described the circumstances in which the searching goes on.

> When a host is under biological control, it has a characteristic 'colonial' or 'spotted' distribution (Smith, 1939). The host population consists of groups of individuals, which fluctuate in density independently of one another. One group may be exterminated by a parasite, but in the meantime some host individuals will have migrated and established new groups. The parasite must therefore follow its host, and so must equal it in powers of locomotion. . . . The female of *Metaphycus helvolus* is very active and apparently travels considerable distances between ovipositions. Observations indicate that populations

of *M. helvolus* may disperse a distance of 25 miles in one year (sections 9.12, 9.2; Figs. 9.03, 9.08).

10.45 *The general importance of dispersal*

Elton (1927) wrote: 'The idea with which we have to start is that animal dispersal is on the whole a rather quiet, humdrum process, and that it is taking place all the time as a result of the normal life of the animals.' If you look again at the examples that I have discussed in this chapter you will see that these animals have, as a normal part of their lives, patterns of behaviour that tend to speed the processes of dispersal and make them more effective. One can often observe in nature circumstances that might be expected to favour the evolution of such patterns of behaviour.

Food and suitable places to live are often scattered 'patchily' and perhaps sparsely through a terrain that is otherwise inhospitable. Moreover the pattern changes from time to time perhaps as a result of ecological succession, perhaps from catastrophe such as flood or drought or fire, or perhaps from the industry of men; or the activities of the animals themselves may speed the process of succession – whatever the cause, places that were suitable for occupation today may have gone by tomorrow, and new ones may have arisen where there was none before. In these circumstances it may be an advantage to a species to breed individuals that readily seek a new place to live, even while the old one is still quite favourable. The passage from Cockerell that I quoted in section 3.12 shows how high powers of dispersal may be an advantage to a species that has to live with a predator that presses heavily upon it, tending to annihilate the local populations that it finds. The following passage which I quote from Nicholson (1947) shows how high 'dispersiveness' is also an advantage to a herbivorous insect that presses heavily on its food-plant, tending to annihilate local populations when it finds them. This passage summarizes the sequence of events that happened after the moth *Cactoblastic cactorum* had been introduced into Queensland with the hope that it would reduce the abundance of prickly pear, *Opuntia*, which was at that time a widespread and serious weed. The venture was a striking success.

After the introduction of this moth into Australia it increased in abundance at a terrific rate, quickly destroying the prickly pear over large areas around the points of liberation. After some time the point was reached at which very little prickly pear remained, whereas there were in the country countless millions of moths that had bred on the prickly pear destroyed during the previous generation. Most of the caterpillars coming from the eggs of these moths died for simple lack of food, and most of the remaining prickly pear was destroyed at the same time. But this did not produce the complete eradication of the prickly pear. By chance, or because of some favourable circumstance, here and there prickly pear plants survived. Similarly, even at this time some few of the moths succeeded in finding odd plants on which to lay their eggs, although because of the great reduction in numbers of *Cactoblastis* in the previous generation, many plants were not found. The end result, which still persists, is that prickly pear is scattered in small isolated groups with wide intervals between them.

Relatively few animals press so heavily on their food as *Cactoblastis* or are pressed on so heavily as *Icerya* (section 5.11); but there are many other causes for temporariness or fluctuation in space of local populations. Animals need to be constantly dispersing as a normal part of their lives – and those that happen to evolve special patterns of behaviour which speed the processes of dispersal are likely to be favoured by natural selection.

10.5 CLASS EXPERIMENTS ON THE ESTIMATION OF DENSITY, DISTRIBUTION AND DISPERSAL

The techniques that are illustrated by the experiments that are described in this section are for use in the field, but most of the experiments are done in the laboratory. With small animals such as *Paramoecium* or *Tribolium* it is practicable to use artificial populations as models of natural ones. Ecology is at its best only when the practical work is done on natural populations in the field; but, as a compromise, the laboratory model of a population is a useful device for introducing the techniques to students who can afford only the limited amount of time prescribed by a rigid time-table.

10.51 *Density*

(See sections 10.1, 10.2)

Experiment 10.511

(*a*) Purpose

To compare the relative densities of two populations of *Paramoecium* in two large petri dishes. This experiment may be regarded as a laboratory model of the field method of 'simple trapping' (section 10.1).

The null hypothesis is that there is no difference in the densities of the two populations.

(*b*) Material

A culture of *Paramoecium* (Appendix).

(*c*) Apparatus

(*i*) Two large petri dishes about 22 cm in dia.

(*ii*) Two sheets of graph paper at least as large as the petri dishes.

(*iii*) Culture medium suitable for *Paramoecium* (Appendix).

(*iv*) Glass sampling rods made by heating one end of a short length of 5-mm glass rod in a flame until the end is slightly bulbous.

(*v*) Microscope slides and a binocular dissecting microscope for counting *Paramoecium*.

(*vi*) A table of random numbers.

(*d*) Method

Place a sheet of graph paper under each petri dish. Half-fill the dishes with medium and transfer an unknown number (preferably several thousand) *Paramoecium* to the dishes.

Take 50 (or thereabout) random samples from each dish. Take the samples by dipping the sampling rod into the medium and then touching the drop of medium on to a microscope slide. Because the null hypothesis is to be tested by comparing the empirical difference between the means with the theoretical distribution of t, it is essential that the samples be taken at random to any known or suspected cause for variation in numbers that is not measured separately. In the present instance to take samples at random to the area of the dish is the best that can be done because we are ignorant of any specific cause for variation. Construct two axes on the graph paper as if to draw a graph. Choose two random numbers from the tables for each sample; use these numbers to plot the position of the sample. Count the *Paramoecium* in each sample and construct a frequency table as in Table 10.511*a*.

The estimates of the mean $\bar{x}$ and the variance s^2 of x for each sample may be calculated from the formulae:

$$\bar{x} = \frac{Sx}{n}$$

$$s^2 = \frac{S(x-\bar{x})^2}{n-1} = \frac{1}{n-1}\left[Sx^2 - \frac{(Sx)^2}{n}\right].$$

Table 10.511*a*

Dish A		Dish B	
Number per drop x_1	Number of drops f	Number per drop x_2	Number of drops f
0	1	0	4
1	1	1	2
2	8	2	4
3	7	3	5
4	5	4	6
5	2	5	1
6	2	6	5
7	4	7	4
8	1	8	2
9	2	9	1
10	1	12	1
13	1	13	1
14	1		
18	1		
19	1		
28	2		

$Sfx_1 = 264$; $n_1(Sf) = 40$; $Sfx_2 = 162$; $n_2(Sf) = 36$

(*e*) Results

If we write x_1 and n_1 for the sample from dish A and x_2 and n_2 for the sample from dish B the estimate of the variance of the difference between the two populations (on the assumption that the variances are homogeneous and can therefore be pooled without bias) is:

$$\text{Variance of difference} = \frac{S(x_1 - \bar{x}_1)^2 + S(x_2 - \bar{x}_2)^2}{n_1 + n_2 - 2}$$

The null hypothesis that the density of the populations in the two dishes do not differ from each other may be tested by calculating t.

$$t = \frac{\bar{x}_1 - \bar{x}_2}{\text{standard deviation of the difference between means}}$$

The standard deviation of the difference between means may be calculated from,

$$\text{S.D. diff.} = \sqrt{\left(\frac{S(x_1 - \bar{x}_1)^2 + S(x_2 - \bar{x}_2)^2}{n_1 + n_2 - 2}\right)} \times \sqrt{\left(\frac{n_1 + n_2}{n_1 \times n_2}\right)}$$

$$= \sqrt{\left(\frac{3{,}438 + 1{,}082}{40 + 36 - 2}\right)} \times \sqrt{\left(\frac{40 + 36}{40 \times 36}\right)}$$

$$= 1{\cdot}796$$

$$t = \frac{6{\cdot}60 - 4{\cdot}50}{1{\cdot}796}$$

$$= 1{\cdot}17$$

t has $n_1 + n_2 - 2$ degrees of freedom; from Fisher and Yates Table III it may be seen that for n 70 and t 1·17, P lies between 0·2 and 0·3.

(*f*) Conclusions and comment

The odds are rather less than 4 to 1 against the null hypothesis. Usually we require the odds against a null hypothesis to be at last 19 to 1 before we accept the difference as 'significant'. Inspection of Table 10.511*a* shows that the distribution, especially the sample taken from dish A, is skewed in the same direction as, but perhaps more strongly than, a Poisson. The t test assumes a small sample from a normal distribution. Even though the original distributions are so skew it is likely that the distribution of the means will be close enough to normal to justify this assumption. So we have failed to disprove our null hypothesis. The conventional way of describing the results of the present experiment would be to say that the difference was non-significant.

This experiment measured the relative, but not the absolute densities of the two populations because we did not know what proportion of the whole medium was contained in the drop that was lifted out with the sampling rod.

The mean number of *Parmoecium* per sample, the standard deviation of the mean and the fiducial limits for the estimate of the mean per sample may be calculated for each dish as in experiment 10.512.

Experiment 10.512

(*a*) Purpose

To estimate the number of flour beetles *Tribolium confusum* in a box of flour. The box is 63·5 cm square and the flour is about four cm deep.

(*b*) Material

A population of flour beetles established and living naturally in the flour.

(*c*) Apparatus

A sampling cylinder made by cutting the ends out of a small can or by rolling a piece of flexible cardboard, a spoon bent at the shank to scoop out the flour, a sieve, a table of random numbers and two pieces of thin dowelling that will comfortably span the box.

(*d*) Method

Mark out parallel scales in units of about 1 cm along the edges of the box. Use these scales as axes of a graph and use the dowelling to plot the positions of the samples. Fix the positions of the samples with numbers drawn from the table of random numbers as in experiment 10.511. Place the sampling cylinder in position. Scoop out and sieve the flour. Count the beetles and replace them and the flour in the box where they came from.

(*e*) Results

In 16 samples we found 936 beetles as follows:

57, 59, 129, 50, 70, 69, 15, 35, 57, 35, 85, 71, 61, 70, 45, 28.

The mean number per quadrat $\bar{x} = \frac{Sx}{n} = 58 \cdot 50$

The variance of x $$s^2 = \frac{1}{n-1}\left[Sx^2 - \frac{(Sx)^2}{n}\right]$$
$$= 693 \cdot 07$$

The standard error of $\bar{x}$ $\frac{s}{\sqrt{n}} = 6 \cdot 582$

The fiducial limits of the mean are calculated:

$$\text{Fiducial limits} = \bar{x} \pm t \times \frac{s}{\sqrt{n}}$$

From Fisher and Yates Table III for $n = 15$, $P = 0 \cdot 05$, $t = 2 \cdot 131$.
So fiducial limits $= 58 \cdot 50 \pm 2 \cdot 131 \times 6 \cdot 582$
$= 72 \cdot 53, 44 \cdot 48$

(*f*) Conclusion and comment

The area of the box was 142·6 times the area of a quadrat so there was one chance in twenty that the population was smaller than 6,343, or larger than 10,343. With 16 quadrats, sampling about one-tenth of the total area, one might have hoped for narrower fiducial limits, i.e. for a more precise estimate. An inspection of the figures for the individual quadrats suggests that the distribution was extremely patchy. If a pattern can be recognized in the patchiness of a distribution it may be possible to improve the accuracy of the estimate, without increasing the labour, by stratifying the sampling as in experiment 10.513.

Experiment 10.513

(*a*) Purpose

To estimate the absolute density of a population of flour beetles *Tribolium confusum* using the method of stratified sampling.

(*b*) Material

The same population of beetles as in experiment 10.512.

(*c*) Apparatus

The same as for experiment 10.512.

(*d*) Method

Because it seemed, by inspecting the tunnels left by the beetles in the flour, that the beetles were more numerous in the corners the area was divided into two equal areas; one embraced the four corners and the other the rest of the box. Samples were taken as in experiment 10.512 except that 12 samples were taken from each area.

(*e*) Results

We found for area I $\bar{x}_1 = 66{\cdot}87$; variance of $\bar{x}_1 = 68{\cdot}29$

for area II $\bar{x}_2 = 40{\cdot}83$; variance of $\bar{x}_2 = 45{\cdot}63$.

So the variance of $(\bar{x}_1 + \bar{x}_2) = 68{\cdot}29 + 45{\cdot}63 = 113{\cdot}92$.

The standard error of $(\bar{x}_1 + \bar{x}_2) = \sqrt{113{\cdot}92} = 10{\cdot}673$

Because the values of n are small (i.e. $n_1 = n_2 < 20$) and the variances have been estimated separately, we may not use the table of t. Instead we use Table VI of Fisher and Yates to find d having first estimated θ from

$$\tan\theta = \frac{s_1^2}{s_2^2} = \frac{45{\cdot}63}{68{\cdot}29} = 0{\cdot}6682$$

So $$\theta = 33^\circ\, 45'$$

And the fiducial limits $= (\bar{x}_1 + \bar{x}_2) \pm d\sqrt{(45{\cdot}63 + 68{\cdot}29)}$

Interpolating in Table VI of Fisher and Yates we find that when $n_1 = n_2 = 11$,

$$P = 0{\cdot}05 \text{ and } \theta = 33^\circ\, 45',\ d \doteqdot 2{\cdot}3$$

So the fiducial limits $= (66{\cdot}87 + 40.83) \pm 2{\cdot}3 \times 10{\cdot}673$

$$= 107{\cdot}70 \pm 24{\cdot}55$$
$$= 132{\cdot}25,\ 83{\cdot}15$$

The area of the half-box is 71·28 times the area of a quadrat.

So there is one chance in twenty that the number of beetles will be less than 5,927 or more than 9,427.

By increasing the number of samples by 50% and using stratified sampling we have slightly increased the accuracy of our estimate; the range of the limits has been reduced from 4,000 to 3,500. If we had taken 20 or more quadrats from each area we might have used the table of t without risk of bias which would probably have given a more accurate estimate.

Experiment 10.514

(*a*) Purpose

To estimate the number of snails *Helix* sp. in a garden by the method of capture, marking, release and recapture (Jackson's negative method).

(*b*) Material

A population of snails living naturally in a garden. The garden was mostly shrubbery. It measured about 150 feet by 50 feet.

(*c*) Apparatus

All that is required is a suitable material for marking the snails. We tried a variety of cellulose and other lacquers which would not stay on the snails during wet weather but we found a proprietary line of nail-polish that was durable. We found six shades that we could distinguish on the recaptured snails.

(*d*) Method

At certain seasons of the year (at Adelaide usually during September, October) the snails may be trapped quite readily under boards or inside inverted flower-pots especially if they are baited with a little metaldehyde. But at other times of the year very few are caught in the traps. Even during the spring their behaviour is erratic and it is better, though more laborious, merely to search for them. Each snail, as it is found, is marked with the colour appropriate to that date and released. The intervals between the days on which the snails are marked and released must be constant. About one week is a convenient interval because it allows the marked snails time to distribute themselves similarly to the unmarked ones, and it does not prolong the experiment unduly. It is desirable to mark and release on at least six occasions; the minimum number that is practicable is three. Some snails may be caught several times. They should be marked with the appropriate mark on each occasion.

(*e*) Results

The results are given in Table 10.514*a*. The value for r_{-}, as calculated from equation 10.5, was 0·8889; the values for y_{-1} to y_{-7} and a were calculated from equations 10.4 and 10.7 respectively. The estimated population on June 19 is given by:

$$P_0 = \frac{10{,}000}{a} = \frac{10{,}000}{2{\cdot}1075} = 4{,}745$$

I mentioned in section 10.23 that it is outside the scope of this book to calculate the variance of this estimate. The fact that we recaptured on

Table 10.514*a*

Date	t_{-7} May 1	t_{-6} May 8	t_{-5} May 15	t_{-4} May 22	t_{-3} May 29	t_{-2} June 5	t_{-1} June 12	t_0 June 19
No. captured and marked	31	57	73	108	97	104	97	
No. recaptured on June 19	4	7	10	15	16	20	23	
Total captured on June 19 (including recaptures)								1,219
y	0·950	1·079	1·119	1·167	1·334	1·607	1·923	
a								2·1075

seven occasions and were able on t_0 to see 95 of the 567 snails (marks) that had been marked on previous occasions suggests that this is likely to be a reasonably good estimate.

Experiment 10.515

(*a*) Purpose

To estimate the trends in the density of a population of *Tribolium castaneum* living in a flat box of flour. It is intended to use the method of capture, marking, release and recapture; Jackson's positive and negative methods will be combined to construct a trellis diagram.

(*b*) Material

A colony of *T. castaneum* containing all stages from eggs to adults. A colony that has been cultured as described in the Appendix and has been living at 27°C in 200 g of flour-medium for eight weeks is suitable.

(*c*) Apparatus

(*i*) A flat box about 75 cm square and 10 cm deep.

(*ii*) Enough flour-medium to make a layer about five cm deep in the box.

(*iii*) A number of glass specimen tubes about 70 mm by 15 mm – for taking samples.

(*iv*) Some white paint or lacquer to mark the beetles.

(*v*) A binocular dissecting microscope.

(*vi*) A cylinder of carbon-dioxide and suitable apparatus for holding the beetles under light anaesthesia while they are marked under the microscope. A buchner funnel mounted in place of the stage of the microscope with CO_2 led into it from below, works quite well.

(*vii*) A table of random numbers.

(*d*) Method

(*i*) About a week before the experiment is due to start tip the culture of *Tribolium* into the box and half-fill the box with flour-medium.

(*ii*) Samples may be taken by pushing a glass specimen tube upside down into the medium and lifting it out. The beetles are sieved out from the medium, counted, marked and scattered in the vicinity of the same place as they came from.

(*iii*) The position of the samples may be randomized by drawing, from a table of random numbers, pairs of numbers to serve as co-ordinates which locate the sample with respect to scales that have been marked along the edges of the box.

(*iv*) The number of samples depends on the number of beetles taken relative to the total population. On the first occasion it may suffice to take enough samples to yield between 100 and 200 beetles. On subsequent occasions it is best to continue sampling until about 5 to 10% of the most recently marked ones have been recaptured.

(*v*) The beetles may be marked, while they are under the anaesthetic, with the extreme tip of a fine sable-hair brush (artists No. 00). There are nine positions that may be used to indicate the date: left and right side of the thorax, anterior, middle and posterior thirds of each elytron, and if a ninth position is needed a spot of a different colour may be put centrally on the thorax.

(*vi*) With nine positions available the marking and releasing may be done on nine occasions. The occasions must be evenly spaced; a week is a convenient interval. Each beetle that is captured must be marked in the position appropriate for that date, irrespective of how many other marks it may be carrying. In counting recaptures it is the marks that are counted.

(*e*) Results

The results of one experiment are set out in Table 10.515*a*. The table is a trellis diagram. The figures in the body of the table are the number of beetles recaptured. Each horizontal row refers to the number marked on one date, as indicated in the left-hand margin, and recaptured on subsequent dates, as indicated along the top. Each vertical column relates to the numbers recaptured on one day, as indicated along the bottom, of those that were marked on preceding days, as indicated in the left-hand margin. Thus each row provides the raw material for estimating the size of the population (on the day of release) by Jackson's positive method; and each column provides the raw data for estimating the size of the population (on the day of recapture) by Jackson's negative method (section 10.23).

Table 10.515*a*. *Crude recaptures. The numbers in each row indicate the number of beetles recaptured on successive dates of those that were marked on the date indicated in the left margin. The columns indicate the numbers recaptured of those that were marked on the dates shown in the left margin* (*negative method*)

Date and number marked		Date recaptured							
		Nov. 3	Nov. 8	Nov. 13	Nov. 18	Nov. 23	Nov. 28	Dec. 3	Dec. 8
Oct. 29	117	6	9	7	9	3	3	6	5
Nov. 3	102		11	3	7	3	5	2	8
Nov. 8	213			10	11	12	6	8	7
Nov. 13	118				6	3	7	0	3
Nov. 18	145					11	6	4	4
Nov. 23	130						9	6	7
Nov. 28	147							6	6
Dec. 3	134								13
Number captured	117	102	215	119	145	131	147	137	179
Date captured	Oct. 29	Nov. 3	Nov. 8	Nov. 13	Nov. 18	Nov. 23	Nov. 28	Dec. 3	Dec. 8

The raw data for recaptures must be corrected to allow for the varying numbers captured and marked on different occasions. This is done by calculating,

$$y_n = R_n \times \frac{100}{C_m} \times \frac{100}{C_n}$$

where R_n is the number recaptured on date n

C_m is the number marked on the date of release

C_n is the total captured (including recaptures) on date n.

The corrected values are shown in Table 10.515*b*. From each row in which there are three or more entries a value for r_+ is calculated and from each column with three or more entries a value for r_-. Equation 10.6 may be used if there are more than four entries, otherwise equation 10.5. Equation 10.7 is then used to calculate a for each row and column, and N is got by dividing 10,000 by a.

In Table 10.515*b* the recaptures that are arranged along any diagonal were all taken at the same interval after release. This consideration may lead to the following argument. Let us imagine that we might have

Table 10.515*b*. *The figures in the body of the table are corrected recaptures* (y) *which have been calculated from the expression,*

$$y_n = R_n \times \frac{100 \times 100}{C_m \quad C_n}$$

The values for R_n, C_m *and* C_n *are given in Table* 10.515*a*

Date marked and released	Date recaptured									N (positive method)
	Oct. 29	Nov. 3	Nov. 8	Nov. 13	Nov. 18	Nov. 23	Nov. 28	Dec. 3	Dec. 8	
Oct. 29		5·0	3·6	5·0	5·3	2·0	1·7	3·7	2·4	1,890
Nov. 3			5·0	2·5	4·7	2·2	3·3	1·4	4·4	(4,333)
Nov. 8				3·9	3·6	4·3	1·9	2·7	1·8	(2,059)
Nov. 13					3·5	1·9	4·0	0	1·4	3,332
Nov. 18						5·8	2·8	2·0	1·5	3,322
Nov. 23							4·7	3·4	3·0	1,801
Nov. 28								3·0	2·3	
Dec. 3									5·4	
N (neg. method)				(6,369)	3,164	2,053	1,955	(4,550)	1,998	
Date	Oct. 29	Nov. 3	Nov. 8	Nov. 13	Nov. 18	Nov. 23	Nov. 28	Dec. 3	Dec. 8	

recaptured the beetles marked and released on, say November 8, immediately after they had been released. Let us call the corrected value of this imaginary recapture y_0; also let us call the corrected value for the actual recapture on November 13 (of beetles marked on November 8) y_1. Similarly let the recaptures on November 8 and 13, of beetles marked on November 3, be called x_1 and x_2 respectively. Now y_1/y_0 is an estimate of r_+ (section 3.23) for the period November 8 to 13; similarly x_2/x_1 is an estimate of r_+ *over the same period*. It is reasonable to equate these two independent estimates of r_+. Thus:

$$\frac{y_1}{y_0} = \frac{x_2}{x_1}$$

$$\therefore y_0 = \frac{x_1 y_1}{x_2}$$

This is generally true in the circumstances in which . . . x, y, z, denote consecutive dates of marking, and $x_1, x_2, x_3 \ldots, y_1, y_2, y_3, \ldots, z_1, z_2, z_3$, denote consecutive dates of recapture. Therefore we have as a general estimate of y_0:

$$a = x_1 \frac{y_1 + y_2 + y_3 \ldots y_n}{x_2 + x_3 + x_4 \ldots x_n} \tag{10.10}$$

This method of estimating a seems to be theoretically sounder than the ones that we have been using because it does not depend so much on extrapolation; but it has the same defect as the others in that it depends on the assumption that the series x and y decline in geometric progressions. This is not likely to be true except in a population in which the likelihood of death is independent of age or else the age-structure of the population is invariable. Jackson (1948) described a method which seems to be more realistic because it does not depend on any of these assumptions.

The estimates of N (the number of adult *Tribolium* in the box) that were made from Jackson's positive and negative methods and from equation 10.10 are brought together in Table 10.515*c*.

(*f*) Conclusion and comment

Table 10.515*c* gives estimates of N for each sampling day from October 29 to December 8. In the absence of any knowledge of the variance of these estimates we cannot tell how precise they are. The occurrence of one or two aberrant values (indicated by parentheses) in each of the three series of estimates suggests that sampling errors may have been large.

Table 10.515c. *Details of the calculations for completing the estimates of the size of the population*

Date	Positive method			Negative method			Equation 10.1	
	r_+	a	N	r_-	a	N	a	N
29 Oct.	0·943	5·290	1,890					
3 Nov.	0·951	2·308	(4,333)				4·958	2,017
8	0·884	4·858	2,059				4·919	2,033
13	0·758	3·001	3,332	1·17	1·570	(6,369)	2·946	3,394
18	0·913	3·010	3,322	1·15	3·161	3,164	5·801	(1,723)
23	0·790	5·552	1,801	0·842	4·872	2,053	10·219	(979)
28				0·902	5·112	1,955	3·892	2,569
3 Dec.				0·940	2·198	(4,550)		
8				0·970	5·006	1,998		

The causes of the fluctuations in N may be inferred from inspection of Table 10.515*b*. The values of y given in the Table really stand for percentages of marked animals recaptured in a standard sample of 100 captures. Consider any row, say the one opposite November 3. If the death-rate between November 3 and December 8 was the same for marked as for unmarked then the proportion of recaptures should remain the same along the full length of the row unless the population has been increased by the addition of unmarked ones by birth or immigration. Consequently any downward trend in the row from left to right may be taken as a measure of the increments due to births and immigration. In this experiment immigration was arbitrarily prevented and births mean the emergence of adults from pupae. Now consider any column, say the one above the recapture date of December 3. If the population has been growing by the addition of unmarked ones through births or immigration this should depress the proportion of marked ones recaptured equally along the whole column. But if the beetles have been dying during the experiment this is likely to depress the figures near the top of the column relatively more than those near the bottom because a larger proportion of the beetles marked early are likely to have died by December 3 than of those marked late. Consequently trends in columns may be taken as a measure of the death-rate in the population.

In Table 10.515*b* there is a clear trend along the rows suggesting that the population was being augmented by births throughout the period of the experiment and especially during the middle of the period. On the other hand the evidence from the columns does not indicate many deaths except perhaps towards the end of the period. The initial increase which we inferred from our estimates of N was associated with an early period when births were occurring in the virtual absence of deaths. Later the population declined because the birth-rate fell and the death-rate increased.

Instead of *Tribolium* the mealworm *Tenebrio molitor* may be used for this experiment. It has the advantage of being large and even an inexperienced person should have no difficulty in putting nine or more distinctive marks on it. The disadvantage is that it develops so slowly. The modifications that are required include a larger box, larger sampling tubes and a bran-medium instead of a flour-medium.

We have also used the snail *Helicella virgata* living in a park.

10.52 *Patterns of distribution*

In section 10.31 I mentioned that an empirical distribution generated by counting all the animals in a number of contiguous quadrats might be compared with a theoretical Poisson distribution with the same mean to test for departures from a random scatter in a natural population. The method has wide application because it can be extended to analogous situations where it may be inconvenient to set up contiguous quadrats or to count every animal in them. The next two experiments illustrate departures from the conventional design.

Experiment 10.521

(*a*) Purpose

To compare the distribution of the scale insect *Saissetia oleae* on an olive bush with the theoretical distribution calculated from the Poisson series. The null hypothesis is the same as in experiment 10.11.

(*b*) Material

Any large shrub or small tree carrying a not too dense population of scale insects. For our exercise we chose a stunted olive bush about two metres high.

(*c*) Apparatus

Secateurs and a basket in which to collect and store the twigs until the scale insects have been counted.

(*d*) Method

Choose some suitable length of twig such that the average twig is likely to have not more than two or three scale insects on it. Clip a convenient number (between 100 and 200) twigs from the bush working systematically over the bush. We took the terminal 10 cm of every branch up to a height of about $1\frac{1}{2}$ m. Count the scale insects on each twig and arrange them in a frequency table as in Table 10.521*a*.

(*e*) Results

The results are presented in Table 10.521*a* which has the same form as Table 10.02, section 10.31. The mean $\bar{x}$ and variance s^2 are estimated from the totals at the foot of the table:

Table 10.521*a*. *The observed frequencies compared with the theoretical frequencies calculated from the Poisson series*

Number of scale insects per twig (x)	Number of twigs (f)	fx	fx^2	Expected frequencies (Poisson)
0	187	0	0	82·69
1	63	63	63	109·54
2	18	36	72	72·55
3	16	48	144	32·04
4	9	36	144	10·61
5	2	10	50	2·81
6	2	12	72	0·62
7	5	35	245	0·12
8	3	24	192	0·02
13	1	13	169	0·00
17	1	17	289	0·00
19	1	19	361	0·00
25	1	25	625	0·00
34	1	34	1,156	0·00
40	1	40	1,600	0·00
Total	311	412	5,182	311·00

$\bar{x} = 1{\cdot}325$; $s^2 = 14{\cdot}96$; ratio $\frac{s^2}{\bar{x}} = 11{\cdot}3$

$$n = Sf = 311$$

$$\bar{x} = \frac{Sfx}{n} = \frac{412}{311} = 1{\cdot}325$$

$$s^2 = \frac{Sf(x-\bar{x})^2}{n-1} = \left[Sfx^2 - \frac{(Sfx)^2}{n}\right] \times \frac{1}{n-1}$$

$$= \left(5182 - \frac{412^2}{311}\right)\frac{1}{310}$$

$$= 14{\cdot}96$$

(f) Comments and conclusions

The discrepancies between the observed and theoretical frequencies in Table 10.521*a* are so large that there is no need for statistics to demonstrate them. The null hypothesis that there is no difference between the two distributions has been disproved at an obviously high (though unmeasured) level of probability. The high value of 11 for the ratio of the variance to the mean shows that the scale insects are distributed patchily. This is also evident from inspection of Table 10.521*a*; there are too many twigs with zero and more than five and too few with one, two or three. If a statistical test were required χ^2 can be estimated:

$$\chi^2_{n-1} = \frac{(n - 1)s^2}{\bar{x}} \text{ (see section 10.31).}$$

Experiment 10.522

(*a*) Purpose

To compare the pattern of distribution in two colonies of *Paramoecium*. We have two null hypotheses. (*i*) That the distribution of the *Paramoecium* in our samples is not significantly different from the Poisson series. (*ii*) That there is no similarity in the distributions of the *Paramoecium* in the two dishes, i.e. the correlation coefficient between them does not differ significantly from zero.

(*b*) Material

A culture of *Paramoecium* (Appendix).

(*c*) Apparatus

(*i*) Two flat glass dishes. We use petri dishes 22 cm in dia.

(*ii*) Two sheets of paper to go under the petri dishes with a square grid ruled on them. The scale of the grid should be such that the bottom of the dish covers about 100 squares.

(*iii*) A sampling rod made by rounding one end of a 5-mm glass rod in a flame.

(*iv*) Microscope slides and a binocular dissecting microscope for counting. *Paramoecium* may be counted under a magnification of 12·5 times.

(*d*) Method

(*i*) Tip medium containing *Paramoecium* into the two petri dishes until it is about half an inch deep. Take a few preliminary samples by dipping the rod into the medium and touching the drop of medium on to a microscope slide. If necessary adjust the density of the population or the sampling rod so that not more than six *Paramoecium* are likely to be taken, on the average, in one sample. Place the dishes on a bench in front of a window in bright light but not in direct sunlight. Place the paper grids under the dishes so that the grids are orientated parallel to each other. Leave the dishes undisturbed for several hours to give the *Paramoecium* time to distribute themselves naturally.

(*ii*) Take a sample from above each square in the grid by dipping the rod into the medium and touching the drop on to a slide. Keep a record of the squares from which the samples were taken so that a table like Table 10.522*a* may be constructed to compare the numbers in corresponding squares in the two dishes.

(*e*) Results

(*i*) The results are given in Table 10.522*a*. The upper numbers in each cell of the table refer to dish A and the lower numbers to dish B. The results from the two dishes have first to be tested against the Poisson series to see whether the *Paramoecium* are distributed at random. If either distribution conforms to the Poisson series there is no point in carrying the analysis any further. But if both distributions are non-random and 'patchy' then it may be appropriate to ask whether the pattern of patchiness in one dish resembles that in the other as it would if both were related to the same external cause such as the position of the bright window or the edge of the dish, etc. On the other hand if the two patterns are not similar we might infer that they have arisen in response to circumstances within the dish – the distribution of the food may be patchy, or perhaps *Paramoecium* is gregarious, forming clusters independent of any external influence.

(*ii*) The frequencies shown in Table 10.522*b* may be constructed from the raw data given in Table 10.522*a*. From the statistics given at the foot of the table it is clear that the distributions are non-random and patchy.

Because n is large P cannot be found directly from the tables of χ^2. Instead we use the table of t having first calculated,

$$t_\infty = \sqrt{(2\chi^2)} - \sqrt{(2n - 1)}$$

But the distribution of χ^2 corresponds to only one tail of the distribution of t. So it is necessary to halve the values of P given in the table of t (Fisher and Yates, 1948, Table III).

Table 10.522*a*. *Numbers of Paramoecium in samples from corresponding positions in two similar dishes*

				4 6	1 10	11 11	5 8	1 4	5 8				
			4 2	2 4	5 5	11 10	7 7	3 2	7 3	9 3			
		1 1	5 5	6 6	10 10	8 8	7 5	11 13	5 10	2 9	5 11		
	4 6	5 7	5 3	5 7	4 4	9 3	7 8	8 9	8 7	4 8	7 7	7 6	
6 3	7 0	7 1	5 6	8 11	6 6	12 19	5 13	9 9	7 4	11 10	9 5	14 1	8 6
4 6	10 4	3 2	11 1	8 7	6 5	7 8	8 16	4 10	4 8	3 0	7 3	5 5	4 8
1 5	12 3	7 2	0 2	10 3	8 3	11 12	13 4	2 6	8 10	2 1	2 10	7 2	8 3
2 6	4 2	6 4	5 3	3 15	5 4	5 7	7 8	5 3	3 5	3 4	4 5	3 3	6 0
10 5	6 3	8 4	5 10	5 10	6 11	5 8	6 8	5 12	9 8	9 6	7 1	6 1	1 6
5 9	10 4	5 3	8 6	11 13	9 4	13 5	4 10	1 5	7 2	6 4	8 7	8 0	5 5
	5 12	6 5	1 2	3 10	6 5	2 3	2 5	7 2	2 4	5 5	5 1	3 7	
		4 16	1 5	2 11	6 7	6 0	3 7	5 5	7 5	4 6	2 8		
			4 4	3 6	1 4	3 3	1 4	2 2	7 2	5 2			
				6 3	0 3	4 1	8 7	0 0	0 5				

(*iii*) The degree of association between the two distributions may be measured by calculating the correlation coefficient:

$$r = \frac{S(x - \bar{x})(y - \bar{y})}{\sqrt{\{S(x - \bar{x})^2 \times S(y - \bar{y})^2\}}}$$

Table 10.522*b*

No. of *Paramoecium* per sample (x)	No. of samples (f)	
	Dish A	Dish B
0	4	6
1	10	9
2	12	13
3	12	19
4	16	17
5	28	21
6	16	15
7	19	12
8	15	13
9	7	4
10	5	12
11	7	5
12	2	3
13	2	3
14	1	0
15	0	1
16	0	2
17	0	0
18	0	0
19	0	1
Total ($Sf = n$)	156	156
$S(x - \bar{x})^2$	1,394·31	2,020·22
$s^2 = \dfrac{S(x - \bar{x})^2}{n - 1}$	8·998	13·034
Mean ($\bar{x}$)	5·62	5·74
$\chi^2_\infty = \dfrac{S(x - \bar{x})^2}{\bar{x}}$	248·6	351·9
Ratio $\dfrac{s^2}{\bar{x}}$	1.6	2·3
P	$<0·001$	$<0·001$

x and y refer to the individual counts from dish A and dish B respectively. If the calculation is to be done on a machine it is convenient to use the identities:

$$S(x - \bar{x})(y - \bar{y}) \equiv Sxy - \frac{Sx.Sy}{n}$$

$$S(x - \bar{x})^2 \equiv Sx^2 - \frac{(Sx)^2}{n}$$

$$S(y - \bar{y})^2 \equiv Sy^2 - \frac{(Sy)^2}{n}$$

In the present instance the numerator,

$$S(x - \bar{x})(y - \bar{y}) = \text{the sum of } 4 \times 6 + 1 \times 10 + 11 \times 11 + \ldots + 8 \times 7 + 0 \times 0 + 0 \times 1$$

$$- \frac{(\text{sum of } 4 + 1 + 11 + \ldots + 8)(\text{sum of } 6 + 10 + 11 + \ldots + 5)}{156}$$

Thus
$$r = \frac{5{,}276 - 5{,}031{\cdot}5}{\sqrt{2{,}224{\cdot}7 \times 1{,}190{\cdot}2}} = 0{\cdot}15$$

There are 154 (i.e. $n - 2$) degrees of freedom for r which is beyond the range of Table VI in Fisher and Yates. So we use the table of t instead, having calculated t from the equation:

$$t_{\infty} = \frac{r\sqrt{(n-2)}}{\sqrt{(1 - r^2)}} = \frac{0{\cdot}15 \times \sqrt{154}}{\sqrt{0{\cdot}85}} = 2{\cdot}018$$

In the bottom line of Fisher and Yates' Table III, P is given as between 0·05 and 0·02 for this value of t.

(*f*) Conclusion and comment

In both dishes A and B the *Paramoecium* were distributed more patchily than might have been reasonably expected from a random scatter. The discrepancies were highly significant in each instance. Consequently it was pertinent to compare the distributions in the two dishes to see whether any external influence (i.e. external to the dish) could be detected that was common to both dishes. As a first step the correlation coefficient was calculated. This turned out to be small, but significant at the 5% level of probability. That is, there was only one chance in 20 that this degree of association would have occurred by chance. Nevertheless the degree of association was slight leading to the conclusion that whatever external stimulus may have been operating its influence was small relative to the internal influences.

The next steps, inquiring into the nature of the 'internal' stimuli and seeking to identify the 'external' stimuli that might be associated with the small but significant positive correlation may be left to the initiative of the class. A number of useful and interesting experiments suggest themselves.

10.53 *The measurement of dispersal*

The two experiments in this section are modelled on Dobzhansky and Wright's (1947) experiment with *Drosophila* (section 10.421). I have used *Paramoecium* for a laboratory model; larvae of *Tribolium* that are nearly fully grown dispersing in a flat box of flour are also quite good for a laboratory model. The snails *Helix aspersa* and *Helicella virgata* have been used successfully for a field experiment. I have tried other species unsuccessfully. Most species that one might think of won't tolerate the intense crowding that they experience at the beginning of the experiment and still behave naturally. Yet if there are not a great many released in the centre at the beginning, the samples taken subsequently in the outer annuli are likely to contain too few animals. This difficulty is inherent in the design of this experiment. But I have not yet come across any other experiment that serves to measure the rate of dispersal quantitatively, as this one does.

Experiment 10.531

(*a*) Purpose

To measure the rate of dispersal of *Paramoecium.*

(*b*) Material

A culture of *Paramoecium* from which 30,000 may be collected and concentrated into about 10 to 15 cm^3 of medium (Appendix).

(*c*) Apparatus

(*i*) A flat dish with a glass bottom. We use one that is 60 cm square and 10 cm deep. For convenience the sides were made of galvanized iron and a sheet of glass was glazed in to make the bottom. But we found that we had to run a layer of beeswax around the edges to keep the medium away from the metal sides. Otherwise small quantities of metal were dissolved in the medium and killed the *Paramoecium*.

(*ii*) Medium made as in the Appendix and strained through a sieve with about 15 mesh to the centimetre.

(*iii*) A sheet of paper a little larger than the floor of the dish marked with a pattern as shown in Fig. 10.531*a*. The diameter of the central circle equals the width of the annuli and each annulus and the central circle are divided into sections of equal area.

(*iv*) A sampling 'spoon'. We use a chemists' deflagrating spoon which has a capacity of about 1 cm³.

(*v*) Microscope slides with small cylinders, 15 mm dia. by 4 mm deep, cut from polythene tubing cemented on to them, with a waterproof cement. We find that two of these rings on one slide will hold a single sample of 1 cm³. Also these rings just fill the field of the 12·5 times objective of the binocular microscope which we find convenient for counting.

(*vi*) A table of random numbers.

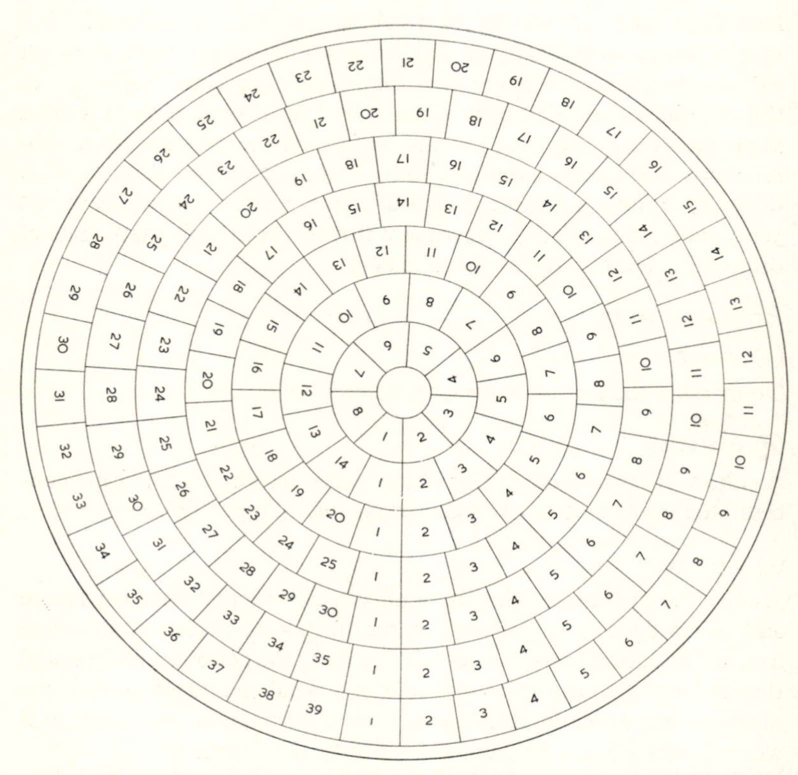

Fig. 10.531*a*. Design of an experiment to measure dispersal. The animals are let go in the central circle. Subsequently samples are taken in the central circle and in each annulus. The diameter of the central circle is equal to the width of an annulus. The annuli are divided into sectors of equal area as an aid to the placing of samples.

(*d*) Method

(*i*) Medium is poured into the dish until it is about 2·5 cm deep.

(*ii*) The *Paramoecium* may be collected and concentrated as follows. Several densely populated cultures of *Paramoecium* in glass jars about 9 cm in dia. and 18 cm deep are placed near a window in bright light but not in direct sunlight. After half an hour or so the *Paramoecium* will have congregated in the top half inch mostly near the edge. They can be drawn off with a fine pipette and collected in small dishes. By further pipetting they may be concentrated until there are about 30,000 in about 12 cm^3. We have found this method satisfactory but there are several other ways of concentrating *Paramoecium* described by Wichterman (1953).

(*iii*) To release the *Paramoecium* place a glass or plastic cylinder about 2·5 cm in dia. by 7 cm long, open at both ends, and ground flat at one end, in the centre of the dish. Hold the cylinder firmly against the bottom of the dish while the *Paramoecium* are gently pipetted into it. Remove the cylinder gently. Note the time.

(*iv*) At the end of one hour take the first lot of samples and thereafter take samples at hourly intervals for six or eight hours. Samples are taken by dipping a 'spoonful' of the medium out and tipping it into the containers for counting. Wash and wipe the spoon thoroughly between samples to avoid spreading the *Paramoecium* artificially – this is important. In the central circle and in the first annulus, especially at first, a spoonful may contain too many *Paramoecium*. So smaller samples may be taken from these places. We use a glass rod as in experiment 10.522. We calibrated the rod and found that the drop collected on the rod was equivalent to 1/25 of a spoonful. So in analysing the results we multiplied the average number per drop by 25 to make them equivalent to the average number per spoon.

(*v*) In order to estimate the density of the population in each annulus with the same precision it would be desirable to take about the same number of animals from each annulus (section 10.22). This would require more samples per unit area in the places where the animals were sparse than where they were dense, and in particular it would require a very large number of samples from the outer annuli in the early stages of the experiment. We did not do this but we compromised by taking samples in proportion to the area of the annulus and we did this by sampling from every third division in each annulus.

The set of divisions to be sampled in each annulus was chosen at random as follows. The divisions in the annuli were numbered as in

Table 10.531*a*. *The number of each annulus is indicated by large Roman numerals. Sf stands for total Paramoecium taken in each annulus and* $\bar{f}$ *stands for the mean number per sample*

No. of hours after release	Number of *Paramoecium* in each annulus																	
	Central circle		I		II		III		IV		V		VI		VII		VIII	
	Sf	$\bar{f}$	*Sf*	$\bar{f}$	*Sf*	$\bar{f}$	*Sf*	$\bar{f}$	*Sf*	$\bar{f}$	*Sf*	$\bar{f}$	*Sf*	$\bar{f}$	*Sf*	$\bar{f}$	*Sf*	$\bar{f}$
1	250	83·33	104	34·67	111	22·20	129	18·43	139	15·44	58	5·27	8	0·72	0	0·00	0	0·00
2	52	17·33	31	10·33	86	17·20	67	9·57	49	5·44	35	3·18	29	2·64	12	0·92	0	0·00
3	51	17·00	16	5·33	18	3·60	23	3·29	26	2·89	28	2·55	18	1·64	12	0·92	7	0·47
4	11	3·67	8	2·67	16	3·20	21	3·00	23	2·56	24	2·18	14	1·27	18	1·38	16	1·07
5	10	3·33	5	1·67	5	1·00	10	1·43	22	2·44	34	3·09	7	0·64	17	1·31	13	0·87
6	8	2·67	5	1·67	6	1·20	5	0·71	26	2·89	20	1·82	12	1·09	20	1·54	15	1·00
No. of samples	3		3		5		7		9		11		11		13		15	

Figure 10.531*a*. For each annulus and for each set of samples we selected a division at random by choosing a number from the table of random numbers. Starting from this division we worked round the annulus clockwise recording the number of every third division. Having thus worked out, in advance, which divisions from each annulus should be used we then arranged all of them (for the whole dish) in a random order, using the table of random numbers, and we sampled from them in this order. In this way we reduced the risk of introducing systematic errors which might have been important if we had sampled the annuli systematically, say from the centre outwards.

(*e*) Results

The results are given in Table 10.531*a*. The first step in estimating the rate of dispersal of the *Paramoecium* is to calculate the variance of the population as indicated by each hourly set of samples. The formula by which this may be done was derived in section 10.421. For example, for the third set of samples,

$$Sr^3\bar{f} = 5{\cdot}33 \times 1 + 3{\cdot}60 \times 8 + 3{\cdot}29 \times 27 + \ldots + 0{\cdot}47 \times 512$$

and

$$Sr\bar{f} + \frac{\bar{c}}{8} = 17{\cdot}00 \times 0{\cdot}125 + 5{\cdot}33 \times 1 + 3{\cdot}60 \times 2 + \ldots + 0{\cdot}47 \times 8$$

The variances for each set of samples, calculated from the expression,

$$s^2 = \frac{Sr^3\bar{f}}{Sr\bar{f} + \dfrac{\bar{c}}{8}}$$

are shown in Table 10.531*b*. They are first calculated in units of d^2 (d is the width of an annulus) and then converted to square centimetres. The standard deviation which is the square root of the variance is, of course, expressed in linear measure.

(*f*) Conclusions and comment

Up to the end of the fourth hour the variance increased fairly steadily. After the fourth hour the variance increased more slowly. This may have been because the *Paramoecium* were artificially restrained by the walls of the vessel in which the experiment was done. At least some of them had reached the walls by this time. So it may be best to base our conclusions on the first four hours.

If we care to assume that the distribution is sufficiently like a normal

Table 10.531*b*. *Variances and standard deviations calculated from the data in Table* 10.531*a*

No. of hrs. after release	Variance in units of d^2 $\dfrac{Sr^3\bar{f}}{Sr\bar{f} + \dfrac{\bar{c}}{8}}$	Variance in cm^2	Hourly increment in variance in cm^2	Standard deviation (cm)
1	10·60	113·70		10·7
2	15·03	161·21	47·5	12·7
3	20·31	217·85	56·6	14·7
4	28·14	301·83	84·0	17·4
5	32·02	343·45	41·6	18·5
6	32·74	351·17	7·7	18·7

distribution we can make a number of inferences from the data in Table 10.531*b*.

(*i*) We know from the table of *t* (Fisher and Yates, 1948, Table III, bottom row) that half the population is likely to be no further from the centre than a distance equal to 0·6745 times the standard deviation. Four hours after the *Paramoecium* were released the standard deviation was 17·4 cm; therefore half the population would be 11·7 cm (0·6745 × 17·4) from the centre. At one hour, half the population was 7·2 cm from the centre. Because, for a normal curve, the median coincides with the mean, this is equivalent to saying that at one hour the average distance of the *Paramoecium* from the centre was 7·2 cm and at four hours the average distance was 11·7 cm.

(*ii*) Between the first and the fourth hours the variance increased on the average by 62·7 cm^2 hr. If we care to assume that, in the absence of artificial barriers, the variance would continue to increase at this rate, then we may predict that by the end of 24 hours the variance would be 1,555·8 cm^2 (113·7 + 23 × 62·7). This gives a standard deviation of 39·4 cm. Multiplying this by 0·6745 we estimate that after 24 hours half the *Paramoecium* will be at least 26·6 cm from the place where they were released.

(*iii*) The table of *t* shows that only 1% of the population in a normal distribution are further away from the centre than a distance equal to 2·5758 times the standard deviation. If we care to treat 1% as negligible then we may say that after 24 hours virtually all the *Paramoecium* will be within 101·5 cm (39·4 × 2·5758) of the place where they were released.

Experiment 10.532

(*a*) Purpose

To measure the rate of dispersal of snails *Helix aspersa* in the field.

(*b*) Material

A convenient number, between 500 and 1,000 snails.

(*c*) Apparatus

Eight-in. flower-pots to serve as 'traps' for the snails. In our experiment we used 119 pots. A tape measure. Some lacquer or paint suitable for marking the snails – we found that nail-polish stayed on best. And a table of random numbers.

(*d*) Method

(*i*) Choose a flat area of ground that is uniformly covered with grass or low herbage but not dissected with flower beds or shrubberies. A circular area about 60 feet in dia. may suffice unless the experiment is done at a time when the snails are unusually active.

(*ii*) Mark out a circle in the centre of the area five feet in dia. Mark out five circles concentric with the central one having radii $7\frac{1}{2}$, $12\frac{1}{2}$, $17\frac{1}{2}$, $22\frac{1}{2}$ and $27\frac{1}{2}$ feet.

(*iii*) Place one flower-pot upside down in the central circle and arrange the remaining 118 pots around the circumferences of the five concentric circles, spacing them four feet apart (or thereabout); but choose the position of the first one in each circle at random as in experiment 10.531. Prop the rim of the flower-pots off the ground so that the snails may crawl into them.

(*iv*) Mark the snails with a spot of varnish and release them in the central circle. We used 600 snails.

(*v*) Next day, and thereafter at daily intervals for the next 7 to 10 days count the marked snails sheltering under each pot and release them near by; then move the pots to a fresh set of random, but equally spaced positions around the same series of circles.

(*vi*) The distances suggested above are suitable for *Helix aspersa* during the winter in Adelaide. With another species or in different weather the snails may be more active and it may be necessary to increase these dimensions.

Table 10.532*a*

No. of days after release	Mean number of snails per trap						Variance s^2 in ft²	S.D. s (ft)	Increase in variance (i)
	Central circle	Annulus							
		I	II	III	IV	V			
1	30	6·500	0·000	0·000	0·000	0·000	15·9	3·99	
2	3	3·500	1·125	0·000	0·000	0·000	51·0	7·14	35·1
3	6	5·375	1·250	0·250	0·000	0·000	59·0	7·68	8·0
4	5	2·875	1·188	0·167	0·065	0·000	79·3	8·91	20·3
5	15	3·625	1·688	0·125	0·098	0·000	69·4	8·33	9·9
6	18	2·625	2·313	0·250	0·226	0·026	101·0	10·05	31·6
7	3	1·875	1·563	0·625	0·194	0·026	143·8	11·99	42·8
8	6	1·375	1·438	0·708	0·258	0·077	170·1	13·04	26·3
9	4	2·375	1·250	0·458	0·129	0·308	203·0	14·25	32·9
10	0	0·500	0·375	0·250	0·290	0·128	294·8	17·17	91·8

(*e*) Results

The results are set out in Table 10.532*a*. The methods of analysis are the same as for experiment 10.531, so the details need not be repeated here.

(*f*) Conclusion and comment

The decrease in variance between the fourth and fifth days may be put down to errors of random sampling. On the ninth and tenth days rather few snails were caught suggesting that by this time many of them had walked right off the area. If we consider the records up to the end of the eighth day we notice a rather large variation in the daily increments of variance. At least some of this was probably due to the influence of weather on the activity of the snails. In a rather more elaborate experiment it might be worthwhile to measure this as Dobzhansky and Wright (1947) did for *Drosophila*. At the end of the eighth day s was 13·04 feet and we may estimate that half the snails were at least 13·04 × 0·6745 feet (= 8·8 feet) from the centre at this time. Between the first and the eight day the variance had increased by 22·0 square feet per day. At this rate the standard deviation on the 30th day would be 24·4 feet, and we may predict that half the snails would be at least 24·4 × 0·6745 feet (= 16·5 feet) away from the centre 30 days after they had been released.

11 *Physiological responses to temperature*

11.1 THE INFLUENCE OF TEMPERATURE ON SPEED OF DEVELOPMENT

See section 6.11.

Experiment 11.11

(*a*) Purpose

To compare the rate of egg-laying by the flour beetles *Tribolium confusum* and *T. castaneum* over a range of temperature. There are three null hypotheses being tested concurrently in this complex experiment.

(*i*) That there is no difference in the number of eggs laid per week (by the two species taken together) at three temperatures 25°C, 27°C, 30°C.

(*ii*) That, taking the three temperatures together, there is no difference in the number of eggs laid per week by *T. confusum* and *T. castaneum.*

(*iii*) If (*i*) and (*ii*) are disproved, that there is no difference between the species with respect to the rate at which the rate of egg-laying changes from one temperature to another.

(*b*) Material

Cultures of *T. confusum* and *T. castaneum* (see Appendix).

(*c*) Apparatus

(*i*) Several incubators. We used 3 that kept the temperature constant to ± 0·2°C at 25°, 27° and 30°C.

(*ii*) Glass specimen tubes (shell vials) about 15 mm wide by 5 cm deep.

(*iii*) Flour-medium (see Appendix).

(*iv*) Sieves made from bolting silk or nylon net of about 12 and 24 mesh to the centimetre.

(*d*) Method

(*i*) *Tribolium* may be readily sexed as pupae but not as adults. So collect enough pupae from the culture for 120 pairs. Place one pair in each of 120 tubes with 8 g of flour-medium. Place the tubes at 30°C 70% R.H. and leave them there for at least three weeks after the last pupa has become adult. Sieve the flour and discard any pair that is not laying eggs.

(*ii*) Divide the 60 pairs of each species randomly (using a table of random numbers) into three lots of 20. Place one lot of 20 pairs of each species at 25°, 27° and 30°C.

(*iii*) At the end of a week's acclimatization sieve the beetles into new flour.

(*iv*) At weekly intervals for the next two weeks sieve the flour; count the eggs; and renew the flour-medium.

(*e*) Results

The mean number of eggs laid per week are given in Table 11.11*a*. *T. castaneum* has laid twice as many eggs at 27° as at 25° but no more at 30° than at 27°, whereas *T. confusum* has laid three times as many eggs at 30° as at 27°C. There seems to be a substantial difference between the species' responses to temperature. An analysis of variance will measure its significance. Because it is more interesting to consider

Table 11.11*a*

Species	Temperature		
	25°C	27°C	30°C
T. confusum	30·4	36·8	119·8
T. castaneum	35·2	74·1	75·6

multiplicative rather than additive differences between the treatments, especially with reference to the interaction between species and temperature, it is appropriate to transform the data to logarithms before carrying out an analysis of variance. The analysis of variance consists essentially of partitioning the total variance into two components, (*i*) that which is associated with differences in temperature or between species and (*ii*) that which is associated with the natural variability between individual beetles receiving the same treatment. If the former

Table 11.11*b*. *Logarithm of mean number of eggs laid per week*

T. confusum						*T. castaneum*					
25° log eggs		27° log eggs		30° log eggs		25° log eggs		27° log eggs		30° log eggs	
x	f	x	f	x	f	x	f	x	f	x	f
1·20	1	0·60	1	1·59	1	1·34	1	1·38	2	1·58	1
1·28	1	1·18	1	1·61	1	1·43	1	1·64	1	1·69	1
1·34	2	1·42	1	1·63	1	1·44	1	1·68	1	1·80	1
1·40	2	1·43	1	1·92	1	1·46	2	1·81	1	1·81	1
1·42	1	1·46	1	1·96	1	1·53	2	1·83	1	1·81	1
1·43	2	1·48	1	2·02	1	1·56	2	1·85	2	1·85	1
1·46	1	1·52	1	2·13	1	1·57	1	1·86	1	1·88	2
1·48	1	1·56	1	2·14	1	1·58	1	1·90	1	1·89	3
1·52	1	1·57	2	2·16	1	1·63	1	1·95	1	1·90	1
1·56	1	1·58	1	2·18	1	1·67	1	1·97	1	1·91	1
1·62	1	1·62	1	2·21	1	1·72	1	2·03	1	1·92	1
1·64	1	1·63	1	2·24	2			2·04	1	1·96	1
1·66	1	1·64	1	2·25	1			2·18	1	2'01	1
1·67	1	1·65	1							2·05	1
		1·72	1								
		1·77	1								
		1·81	1								
$\bar{x}$ 1·46		1·51		2·02		1·53		1·82		1·87	
Sf 17		18		14		14		15		17	
Sfx 24·85		27·21		28·28		21·48		27·35		31·72	
Sfx^2 36·620		42·367		57·902		33·092		50·579		59·381	
$S(x-\bar{x})^2$ 0·295		1·235		0·776		0·136		0·711		0·195	

exceeds the latter by more than a certain ratio we accept the differences associated with the treatments as 'significant'. Moreover, the influence of the treatments may be analysed into three components: (*i*) that due to differences between the same species at different temperatures, (*ii*) that due to differences between the two species at the same temperature, and (*iii*) that due to differential responses of the two species to changes in temperature. The method and the theory of the analysis of variance are discussed by Moroney (1953, pp. 233, 371).

The detailed results, transformed to logarithms, are given in Table 11.11*b*. The figures in Table 11.11*b* may be used to construct an 'interaction table' (Table 11.11*c*).

Table 11.11*c*

	25°		27°		30°		Total	
	Sfx	*Sf*	*Sfx*	*Sf*	*Sfx*	*Sf*	*Sfx*	*Sf*
T. confusum	24·85	17	27·21	18	28·28	14	80·34	49
T. castaneum	21·48	14	27·35	15	31·72	17	80·55	46
Total	46·33	31	54·56	33	60·00	31	160·89	95

The analysis proceeds as follows:

$$\text{Correction factor C.F.} = \frac{(Sx)^2}{n} = \frac{160{\cdot}89^2}{95} = 272{\cdot}480$$

$$\text{Total sum of squares} = Sfx^2 - \text{C.F.}$$

$$= 36{\cdot}620 + 42{\cdot}367 + \ldots + 59{\cdot}381 - 272{\cdot}480 = 7{\cdot}461$$

$$\text{The sum of squares for treatments} = S\frac{(Sfx)^2}{n} - \text{C.F.}$$

$$= \frac{24{\cdot}85^2}{17} + \frac{21{\cdot}48^2}{14} + \ldots + \frac{31{\cdot}72^2}{17} - 272{\cdot}480 = 4{\cdot}113$$

$$\text{The sum of squares for temperatures} = S\frac{(Sfx)^2}{n} - \text{C.F.}$$

$$= \frac{46{\cdot}33^2}{31} + \frac{54{\cdot}56^2}{33} + \frac{60{\cdot}00^2}{31} - 272{\cdot}480 = 3{\cdot}096$$

The sum of squares for species $= S\frac{(Sfx)^2}{n} - \text{C.F.}$

$$= \frac{80{\cdot}34^2}{49} + \frac{80{\cdot}55^2}{46} - 272{\cdot}480$$

$$= 0{\cdot}295$$

The sum of squares for interaction (species × temperature) is got by subtracting the sum of squares for species plus the sum of squares for temperature from the sum of squares for treatments.
Thus:

The sum of squares for interaction $= 4{\cdot}113 - (3{\cdot}096 + 0{\cdot}295)$
$= 0{\cdot}722$

The residual sum of squares can be got by subtracting the sum of squares for treatments from the total sum of squares. Thus:

$$\text{Residual sum of squares} = 7{\cdot}461 - 4{\cdot}113 = 3{\cdot}348$$

The last item can be cross checked by summing the sums of squares for the individual treatments that are given at the foot of Table 11.11*b*:

$$0{\cdot}295 + 1{\cdot}235 + 0{\cdot}776 + 0{\cdot}136 + 0{\cdot}711 + 0{\cdot}195 = 3{\cdot}348$$

The remainder of the analysis may be carried out in Table 11.11*d*. The variance ratio is got by making a ratio of appropriate pairs of mean squares using the one with the larger degrees of freedom as the denominator. The probability is got from Table V of Fisher and Yates (1948). The interaction mean square is first compared with the residual mean square. The interaction is highly significant ($P < 0{\cdot}01$). Because the interaction is significant there is little point in comparing the main effects with the residual mean square. The mean square for temperature is not significantly greater than the mean square for interaction (variance

Table 11.11*d*

Source of variance	Degrees of freedom	Sums of squares	Mean square	Variance ratio	*P*
Total	94	7·461			
All treatments	5	4·113	0·823	21·89	<0·001
Species	1	0·295	0·294		
Temperature	2	3·096	1·548		
Interaction (spec. × temp.)	2	0·722	0·361	9·60	<0·001
Residual	89	3·348	0·0376		

ratio $= \frac{1{\cdot}548}{0{\cdot}361} = 4{\cdot}29; P > 0{\cdot}1$), so there is no evidence that the species differ in their response to temperature except by virtue of the interaction.

(*f*) Conclusions

The most important conclusion to be drawn from this experiment is that the species seem to respond differently to this range of temperature. Whereas *T. castaneum* is responsive to a change of temperature in the lower part of the range *T. confusum* is most responsive at the higher end of the range. Compare the results of this experiment with Fig. 6.111 and the discussion in section 6.02.

In experiment 11.12 we shall use the technique of linear regression to measure the response of an insect to temperature over a wider range. In experiment 12.23 we shall use the technique of multilinear regression to measure the independent influence of two interacting variables.

Experiment 11.12

(*a*) Purpose

To measure the influence of temperature on the rate of egg-production by *Thrips imaginis*. There are two null hypotheses implicit in the design of this experiment.

(*i*) That there is no difference in the number of eggs laid per day by *Thrips imaginis* at the different temperatures.

(*ii*) That the daily number of eggs observed were no different from the numbers that would have been predicted, by drawing a straight line through the observations by the method of 'linear regression'. I have not troubled to calculate the precise odds against the first one because the significance of the results is obvious.

(*b*) Material

Thrips imaginis is indigenous to Australia. But this experiment could be done with any suitable local species of flower-inhabiting thrips (Terebrantia). It is sufficient to breed or collect from the field about 60 to 100 females. Males will probably not be needed because most species of thrips lay eggs quite readily without being fertilized.

(*c*) Apparatus

(*i*) Several incubators that will keep the temperature constant within the range of about 13° to 25°C. We used five set at 10°, 13°, 16°, 20° and 23°C.

(*ii*) A number of small glass tubes about 5 mm by 50 mm.

(*iii*) A fine camel-hair brush for handling the thrips.

(*iv*) Some stamens of *Antirrhinum*, *Pentstemon* or some other fleshy stamen suitable as food for the thrips and for them to lay their eggs into.

(*v*) Microscope slides, gum chloral mounting medium (Berleses fluid or its equivalent) and a microscope for counting the eggs.

(*d*) Method

(*i*) Into each of, say, 100 tubes place one female thrips together with a fresh stamen. Take care to see that the anther is attached to the stamen because the thrips may not lay eggs without pollen. The thrips may be lightly anaesthetized with CO_2 if they are difficult to handle or to sex without anaesthetic.

(*ii*) Divide the hundred tubes at random into five lots, and place each lot of tubes in an incubator in a closed jar (or desiccator) over a saturated solution of NaCl. This will maintain the relative humidity at about 78%. We found the temperatures 10°, 13°, 16°, 20° and 23°C suitable for *Thrips imaginis*, but other species may require a different range of temperature.

(*iii*) Every day for 10 to 15 days remove the old stamen and replace it with a fresh one, taking care not to harm or to lose the thrips.

(*iv*) Mount the old stamen under a coverslip on a microscope slide. Heat the mount gently over a small flame until the medium nearly boils. This will clear the plant-tissue but leave the eggs slightly opaque so that they can be easily counted under a microscope.

(*e*) Results

The results of our experiment with *Thrips imaginis* are set out in Table 11.12*a*. We kept the experiment going for 10 days and the table

Table 11.12*a*. *The mean daily number of eggs laid by Thrips imaginis at different constant temperatures*

Temp. °C	No. of thrips	Mean daily number of eggs laid by each thrips during a 10-day period										Total	Mean
10	10	1·2	0·9	1·6	1·4	1·5	0·8	1·4	1·1	1·1	1·0	12·0	1·20
13	8	2·5	1·9	2·0	2·2	1·7	3·1	2·1	2·3			17·8	2·23
16	10	5·8	4·1	3·6	4·2	4·8	5·4	5·3	3·9	6·2	4·7	48·0	4·80
20	9	7·9	7·1	6·2	8·1	8·7	8·3	9·1	10·5	8·7		74·6	8·29
23	8	8·7	7·6	8·9	6·6	8·7	11·2	8·5	11·0			71·2	8·90

shows the average daily production of eggs for each of the thrips that were still alive at the end of the 10 days. That is why there are unequal numbers of thrips at different temperatures.

The relationship between temperature and rate of egg-production may be seen in Fig. 11.12*a*. As the temperature rose from 10° to 23°C the rate increased from 1·2 to 8·9 eggs per thrips per day. The average increment in daily egg-production for unit increase in temperature over this range may be estimated by calculating the coefficient for linear regression.

$$b = \frac{S(x - \bar{x})(y - \bar{y})}{S(x - \bar{x})^2}$$

where x stands for temperature and y for number of eggs.

If the calculation is to be done on a machine it is best to write

$$S(x - \bar{x})(y - \bar{y}) \text{ as } Syx - \frac{Sx\,.\,Sy}{n} \text{ and } S(x - \bar{x})^2 \text{ as } Sx^2 - \frac{(Sx)^2}{n}.$$

There are 45 pairs of values x, y and therefore $n = 45$.

$$\begin{aligned} b &= \frac{4{,}249{\cdot}000 - 3{,}617{\cdot}351}{12{,}744{\cdot}000 - 11{,}777{\cdot}422} \\ &= \frac{631{\cdot}649}{966{\cdot}578} \\ &= 0{\cdot}6535 \text{ eggs per thrips per } 1^\circ\text{C} \end{aligned}$$

The theoretical relationship between temperature and daily rate of egg-production (assuming a steady increment in egg-production through the range 10°–23°C) may be expressed by the equation

$$\begin{aligned} Y &= \bar{y} + b(x - \bar{x}) \\ &= 4{\cdot}9689 + 0{\cdot}6535\,(x - 16{\cdot}1778) \\ &= -\,5{\cdot}6033 + 0{\cdot}6535x \end{aligned}$$

The straight line that corresponds to this equation is plotted in Fig. 11.12*a*. The observed points do not follow the straight line very closely and the freehand curve (broken line) that has been drawn through them suggests that a sigmoid curve may have been better. This suggestion would be justified if we could show that the discrepancies between the observed points and the theoretical straight line are greater than might reasonably be expected by chance from errors of random sampling. This test is made by calculating χ^2, which is the ratio of two variances. In the numerator we put the variance that is due to systematic departures of the samples (as indicated by their means) away from the theoretical line. And in the denominator we put the variance due to the natural variability of the individual thrips in each sample.

Thus:

$$\chi^2 = S\frac{(Y - \bar{y})^2}{s^2}$$

where Y is the theoretical mean rate of egg-production as indicated by

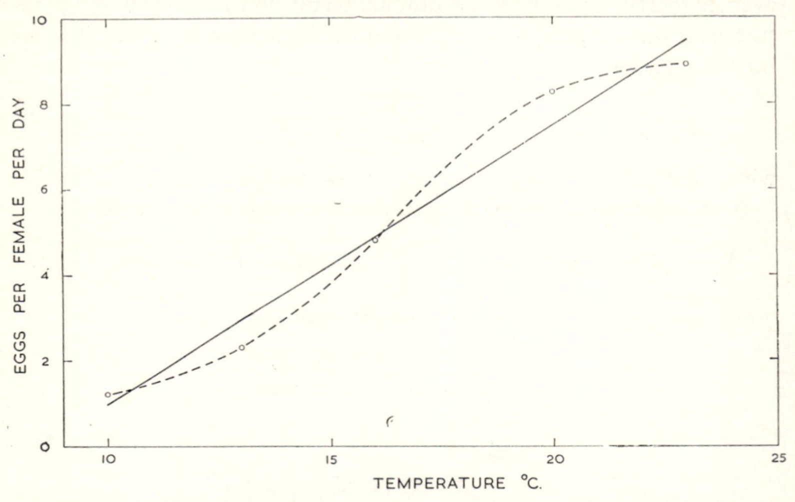

Fig. 11.12*a*. The influence of temperature on the rate of egg production by *Thrips imaginis*.

the straight line, $\bar{y}$ is the actual mean rate of egg production as found by experiment and s^2 is the variance of $\bar{y}$

$$s^2 = \frac{S(y - \bar{y})^2}{n(n - 1)}$$

The calculation of χ^2 is set out (step by step) in Table 11.12*b*. There are three degrees of freedom for χ^2 and Table IV of Fisher and Yates (1948) shows that the probability of such a large value for χ^2 arising from errors of random sampling is less than 1 in 1,000.

(*f*) Conclusion and comment

The highly significant value for χ^2 disproves the null hypothesis at long odds. We conclude that the straight line does not adequately represent the relationship between temperature and rate of egg-production. The trend seems to be sigmoid rather than linear and this suggests that the

Table 11.12*b*. *Calculation of χ^2 to test the goodness of fit of the straight line fitted to the data from Table 11.12a*

Temp. °C	Number of thrips	Sy^2	$\frac{(Sy)^2}{n}$	$S(y-\bar{y})^2$	s^2	$(Y-\bar{y})^2$	$\frac{(Y-\bar{y})^2}{s^2}$
10	10	15·04	14·40	0·64	0·007	0·072	10·29
13	8	40·90	39·61	1·29	0·023	0·438	19·04
16	10	237·08	230·40	6·68	0·074	0·003	0·04
20	9	630·20	618·35	11·85	0·165	0·677	4·10
23	8	650·60	633·68	16·92	0·302	0·278	0·92

Total of last column $= \chi_3^2 = 34{\cdot}4$ $\quad P < 0{\cdot}001$

logistic equation may fit the data more closely (sec. 6.111). The method for calculating the constants of the logistic equation was described by Davidson (1944); and Browning (1952) described the method for testing the goodness of fit.

11.2 THE LETHAL INFLUENCE OF EXTREME TEMPERATURE

The method of 'biological assay' was first used to measure the responses of animals to drugs. It has now become the standard technique for testing insecticides and other poisons. It is the appropriate method for ecologists to use when measuring the 'tolerance' of animals to heat, cold, dryness and other harmful components of environment. The statistical method of 'probit analysis' which has been especially designed for analysing the results of such experiments has been discussed by Finney (1947). He described the theory of the method and worked through a number of examples which are very helpful to the beginner. (See section 6.12.)

Experiment 11.21

(*a*) Purpose

To estimate the duration of exposure to 5°C that is required to kill half the population of *Calandra granaria*, which is also the best estimate that we can get of the mean length of life of *C. granaria* at 5°C (sec. 6.12). The null hypothesis is that there is no difference in the survival-time for *C. granaria* exposed to 5°C and those not so exposed but left at a favourable temperature (27°C in this experiment).

(*b*) Material

About 600 young adults of *Calandra granaria*. Weevils between the ages of 0 and 7 days may be got by sieving the adults out of a culture, leaving it in the incubator for seven days and then sieving it again to collect the weevils that have emerged since the previous sieving.

(*c*) Apparatus

(*i*) Constant temperature cabinets set to run at 5°C and 27°C.

(*ii*) Six small screw-cap jars of about 20 cm^3 capacity, with a ventilator of fine gauze built into the lid.

(*iii*) Some weevil-free wheat.

(*d*) Method

Count 100 weevils at random into each of the six jars and add a little weevil-free wheat. Place one jar, chosen at random from the six in the incubator at 27°C. Place the other five jars at 5°C. At intervals of 4, 7, 11, 14 and 18 days take one jar at random and place it at 27°C. After each jar has been at 27°C for one week, sieve the wheat and record the number of weevils dead and alive.

(*e*) Results

The results of a typical experiment are set out in Table 11.21*a*. This table and the analysis of the results are modelled on the example given by Finney (1947, p. 25 *et seq.*).

The first four columns in Table 11.21*a* are self-explanatory. The fifth column is derived from the fourth by correcting for the deaths that occurred in the controls. The correction is calculated by Abbot's formula as follows.

If p' is the total death-rate, c that part of the death-rate not associated with the treatment and p the death-rate caused by the treatment then,

$$p' = c + p\,(1 - c)$$

and

$$p = \frac{p' - c}{1 - c}$$

For present purposes the death-rate in the control may be taken as an adequate estimate of c.

The last two columns give the data from which Fig. 11.21*a* was drawn. The 'dose' is transformed to logarithms because the probits fall more closely along a straight line when this is done. The probits are derived from the percentages given in column 5; they may be read from Table I of Finney (1947) or from Table IX of Fisher and Yates (1948).

Table 11.21*a*. *The death-rate in samples of Calandra granaria exposed for various periods to 5°C*

Days at 5°C	No. of weevils (n)	No. dead (r)	Proportion dead (p')	Estimated proportion killed by cold (p)	Log days (x)	Empirical probit
0	95	6	0·063	0		
4	102	10	0·098	0·037	0·60	3·21
7	94	25	0·266	0·216	0·85	4·22
11	102	37	0·363	0·319	1·04	4·53
14	99	65	0·657	0·633	1·15	5·34
18	99	82	0·828	0·817	1·26	5·90

The next step is to test the 'goodness of fit' of the provisional line that we have drawn in Fig. 11.21*a* and we do this in Table 11.21*b*. The 'theoretical probits' are read from the graph and entered in Table 11.21*b* under the heading Y. In Table 11.21*b* x, n and r have the same meanings as in Table 11.21*a*; P is the proportion corresponding to

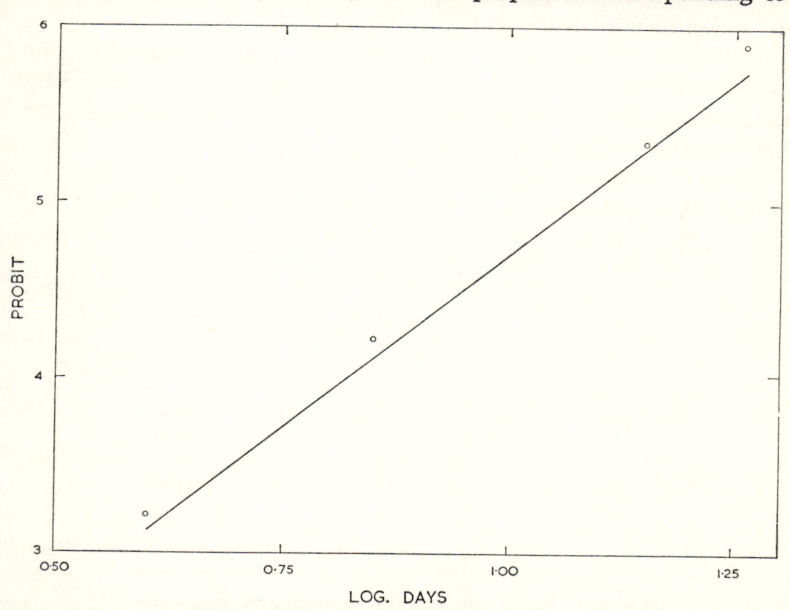

Fig. 11.21*a*. Probit-line for *Calandra granaria*. The abscissae are the logarithms of the number of days at 5°C.

Table 11.21*b*. *Test for 'goodness of fit' of the provisional probit line in Fig. 11.21a*

x	Y	P	P'	n	r	nP'	$r - nP'$	$\frac{(r - nP')^2}{nP'(1 - P')}$
0·60	3·12	0·03	0·091	102	10	9·3	0·7	0·06
0·85	4·12	0·19	0·241	94	25	22·7	2·3	0·32
1·04	4·85	0·44	0·485	102	37	48·5	11·5	5·15
1·15	5·30	0·62	0·641	99	65	63·5	1·5	0·10
1·26	5·74	0·77	0·784	99	82	77·6	4·4	1·14

$\chi^2_3 = 6{\cdot}77$ which is non-significant

Y, *P'* is derived from *P* by applying Abbot's formula, *nP'* is the number of deaths predicted by the graph after allowing for the deaths in the control, $r - nP'$ is the discrepancy between the observed and the predicted number of deaths; and the sum of the quantities in the last column gives an approximate estimte of χ^2. The transformation from *P* to *P'* is necessary because *r* includes deaths from causes other than the treatment and therefore the appropriate comparison is between *r* and *nP'* rather than between *r* and *nP*.

In the present instance χ^2 is non-significant so we may complete our estimate of L.D. 50 and its reliability with the statistics provided by the provisional line of Fig. 11.21*a*, without making any correction for heterogeneity.* From the graph:

When $Y = 5$, $x = 1{\cdot}07$

Also, for unit increase in *x*, *Y* increases by 3·98.

Table 11.21*c*. *Estimate of variance of m*

x	n	y	w $c = 0{\cdot}06$	nw	nwx	nwx^2
0·60	102	3·12	0·051	5·202	3·121	1·873
0·85	94	4·12	0·351	32·994	28·045	23·838
1·04	102	4·85	0·551	56·202	58·450	60·788
1·15	99	5·30	0·558	55·242	63·528	73·057
1·26	99	5·74	0·476	47·124	59·376	74·814
Total				196·764	212·520	234·370

* If χ^2 were significant, the correction for heterogeneity consists in multiplying the variance by the ratio of χ^2 to its degrees of freedom, i.e. heterogeneity factor $= \chi^2/n$.

So we can write,

$$m = 1{\cdot}07;\ b = 3{\cdot}98$$

Taking the antilog of m we get

$$\text{L.D. } 50 = 11{\cdot}7 \text{ days}$$

In order to discover the reliability of our estimate of m we must complete the calculations shown in Table 11.21*c*. The first three columns are copied from Table 11.21*b*; w is the weighting coefficient corresponding to Y which can be read from Table II of Finney (1947). The products nw, nwx and nwx^2 are got by multipying n by w, nw by x and nwx by x.

The variance of m is given by

$$V_m = \frac{1}{b^2}\left(\frac{1}{Snw} + \frac{(m - \bar{x})^2}{Snw\,(x - \bar{x})^2}\right)$$

$$\bar{x} = \frac{Snwx}{Snw} = \frac{212{\cdot}520}{196{\cdot}764} = 1{\cdot}080$$

$$Snw\,(x - \bar{x})^2 = 234{\cdot}370 - \frac{212{\cdot}520^2}{196{\cdot}764}$$

$$= 4{\cdot}832$$

Therefore

$$V_m = \frac{1}{3{\cdot}98^2}\left[\frac{1}{196{\cdot}74} + \frac{(1{\cdot}07 - 1{\cdot}08)^2}{4{\cdot}832}\right].$$

$$= 0{\cdot}000321 + 0{\cdot}0000013$$

$$= 0{\cdot}000322$$

The standard deviation of m is given by:

$$S.D._m = \sqrt{V_m}$$

$$= 0{\cdot}018$$

The fiducial limits of m are given by:

Fiducial limits of m $= m \pm t \times S.D._m$

For 5% probability t $= 1{\cdot}96$

Fiducial limits $= 1{\cdot}07 \pm 1{\cdot}96 \times 0{\cdot}018$

$= 1{\cdot}105,\ 1{\cdot}035$

Taking antilogs we get,

Fiducial limits of $\text{L.D.}_{50} = 12{\cdot}7,\ 10{\cdot}8$ days.

(*f*) Conclusions and comment

We have estimated the L.D. 50 as 11·7 days. And the reliability of our estimate is such that if we were to repeat the experiment there is a 5% chance of getting an estimate that is either less than 10·8 or greater than 12·7 days. The L.D. 50 is the 'dose' of cold that is required to kill half the population. As was explained in section 6.12 the L.D. 50 may also be regarded as an estimate of the mean duration of life of *Calandra granaria* at 5°C.

We did not measure the survival-time of the controls at 27°C but since 94% of them were still alive at the end of the experiment the null hypothesis has clearly been disproved. But in this instance the disproving of the null hypothesis is less important than the estimation of the L.D. 50 and its fiducial limits.

This experiment was analysed by the approximate method of Finney. The graph, Fig. 11.21*a*, is what Finney calls a 'provisional probit line' because it was merely drawn by eye through the provisional probits. For the simple problem of estimating the L.D. 50 and its fiducial limits the approximate method is often sufficient; and it saves a lot of work. Experiment 11.22 is more subtle and we shall have to use a more precise method of analysis, which is equally straightforward but more laborious.

Experiment 11.22

(*a*) Purpose

To estimate the duration of exposure to 5°C that is required to kill half the population of *Tribolium confusum* and *T. castaneum* in the pupal stage. And to test the null hypothesis 'that there is no difference between the L.D. 50 at 5°C for the two species'. The null hypothesis that is implicit in the first part of the experiment is that the survival-time for *Tribolium* that have been exposed to 5°C is no different from that of *Tribolium* that have been kept at a favourable temperature (27°C).

(*b*) Material

About 350 newly moulted pupae of each species.

(*c*) Apparatus

(*i*) Constant temperature cabinets running at 5° and 27°C.

(*ii*) Glass specimen tubes about 6 × 2·5 cm.

(*iii*) A desiccator or some other suitable glass vessel in which the relative humidity is maintained at about 70%.

(*d*) Method

(*i*) In order to get freshly moulted pupae collect rather more (perhaps twice as many) large, apparently fully-fed larvae than the number of pupae that are required. Place them in flour-medium at 27°C. Each day collect the pupae from the medium and replace the larvae. The number of pupae collected each day should be divided equally, and at random, between all the treatments in the experiment. This should be repeated each day until the total number of pupae in each treatment approaches 50.

(*ii*) The pupae are placed in glass tubes with a little flour-medium. From each replicate one lot (the control) is placed directly at 27°C 70% R.H. The others are placed at 5°C to be removed after 2, 3, 5, 7, 9 and 11 days and placed with the controls at 27°C. They are left at 27°C until all the pupae have either emerged as adults or died. But they should be examined frequently and the adults removed as they appear. Otherwise they may eat those that are still pupae.

(*e*) Results

The results of a typical experiment are set out in Table 11.22*a*. The symbols have similar meanings to those in experiment 11.21.

Figure 11.22*a* is drawn from the data in columns 1 and 5. In this instance there is no need to transform days to logarithms because the relationship between empirical probits and days is already sufficiently linear.

One of the purposes of this experiment is to measure the significance of the difference between the L.D. 50 for the two species. We estimate the variance of the difference by pooling the variances of the estimates of L.D. 50 for both species. This implies that the two probit lines have the same slope. So we start by drawing the two provisional probit lines parallel. But before we can proceed to test the significance of the difference in L.D. 50 we must first test the lines to see if they really are parallel. If they differ significantly in slope then the device of drawing them parallel would result in an underestimation of the variance of the L.D. 50 and it would be necessary to increase the estimated variance by a 'heterogeneity factor'. For this test the approximate method of experiment 11.21 is not sufficient; it is necessary to calculate the probit regression equation by the method of 'maximum likelihood'. In practice this consists of starting with the provisional line, drawn by eye as before, and using the statistics from this line to calculate the regression coefficient by the maximum likelihood equations.

Therefore we draw, through each set of empirical probits in Figure 11.22*a*, what we judge by eye to be the best fitting straight lines subject

Table 11.22*a*. *The death-rate in Tribolium confusum and T. castaneum exposed to 5°C*

T. confusum

x	n	r	p (c=0)	Emp. probit	Y	y	w	nw	nwx	nwy
0	39	0	0							
3	39	3	0·08	3·54	3·45	3·60	0·253	13·767	41·301	49·561
5	39	11	0·28	4·43	4·25	4·43	0·518	20·202	101·010	89·495
7	39	21	0·54	5·09	5·04	5·10	0·636	24·804	173·628	126·500
9	39	27	0·69	5·50	5·82	5·46	0·497	19·383	174·447	105·831
Total								78·156	490·386	371·387

T. castaneum

x	n	r	p (c=0·04)	Emp. probit	Y	y	w	nw	nwx	nwy
0	25	1	0							
3·5	25	5	0·17	4·05	4·27	4·06	0·444	11·100	38·850	45·066
5·5	25	12	0·46	4·90	5·08	4·90	0·589	14·725	80·988	72·153
7	25	20	0·79	5·81	5·67	5·80	0·511	12·775	89·425	74·095
10	25	24	0·96	6·75	6·30	6·63	0·321	8·025	80·250	53·208
Total								46·625	289·513	244·522

x = the duration of exposure to 5° in days
n = the number of pupae in each treatment
r = the number that died as pupae
p = the proportion dead, has been corrected by Abbot's formula for *T. castaneum* but not for *T. confusum* because none died in the control
Y = the theoretical probit is read from the graph in Figure 11.22*a*
y = the working probit is taken from Table IV of Finney (1947)
w = the weighting coefficient is taken from Table II of Finney (1947)
nw, nwx and nwy are got by multiplying n, w, x and y together in the combinations indicated by the headings to the columns.

only to the restriction that they must be parallel. Reading from these lines we may write the equations:

For *T. confusum* $Y = 2{\cdot}27 + 0{\cdot}39x$
For *T. castaneum* $Y = 2{\cdot}91 + 0{\cdot}39x$

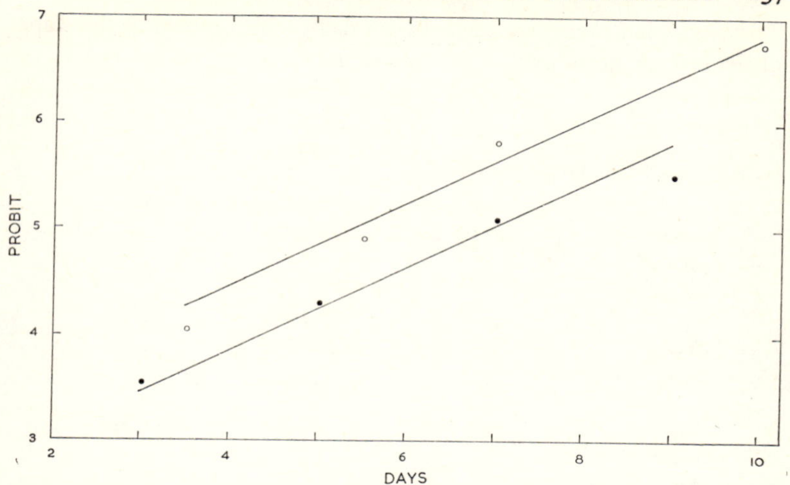

Fig. 11.22*a*. Probit-lines for *T. castaneum* (*hollow circles*) and *T. confusum* (*solid circles*). The abscissae are days at 5°C. The lines have arbitrarily been drawn parallel.

The values for Y in Table 11.22*a* may be calculated from these equations or read directly from the graph. From the data in Table 11.22*a* we calculate:

$Snwx^2$ by multiplying each value of nwx by x and summing the products

$Snwy^2$ by multiplying each value of nwy by y and summing the products

$Snwxy$ by multiplying each value of nwx by the corresponding value of y and summing the products

Then

$$Snw\,(x - \bar{x})^2 = Snwx^2 - \frac{(Snwx)^2}{Snw}$$

written for convenience as S_{xx})

$$Snw\,(y - \bar{y})^2 = Snwy^2 - \frac{(Snwy)^2}{Snw}$$

(written as S_{yy})

$$Snw\,(x - \bar{x})(y - \bar{y}) = Snwxy - \frac{Snwx \times Snwy}{Snw}$$

(written as S_{xy})

The regression coefficient of probit on days, which measures the slope of the lines, is given by:

$$b = \frac{S_{xy}}{S_{xx}}$$

In the present instance we have:

	T. confusum	*T. castaneum*	Total
$Snwx^2$	3,414·372	2,009·884	
$\frac{(Snwx)^2}{Snw}$	3,076·903	1,797·700	
S_{xx}	337·469	212·184	549·653
$Snwxy$	2,434·137	1,605·295	
$\frac{Snwx \times Snwy}{Snw}$	2,330·250	1,518·333	
S_{xy}	103·887	86·962	190·849
$\frac{Snwx}{Snw} = x$	6·274	6·209	
$\frac{Snwy}{Snw} = y$	4·752	5·244	

$$b \text{ (calculated from totals for both species)} = \frac{190{\cdot}849}{549{\cdot}653} = 0{\cdot}347$$

The lines with this slope pass through $\bar{x}$ and $\bar{y}$. Therefore we can write the equations:

T. confusum $\quad Y = 4{\cdot}752 + 0{\cdot}347(x - 6{\cdot}274)$
$\quad = 2{\cdot}575 + 0{\cdot}347x$

T. castaneum $\quad Y = 5{\cdot}244 + 0{\cdot}347(x - 6{\cdot}209)$
$\quad = 3{\cdot}089 + 0{\cdot}347x$

The difference between 0·39 and 0·35 (about 13%) in the two estimates of b is rather large, indicating that we did not judge the slope of the provisional lines (Fig. 11.22*a*) accurately enough. It will therefore be necessary to carry the maximum likelihood calculations through one more step in order to approach more closely to the best fitting lines. Table 11.22*b* is drawn up using the values of Y calculated from the equations given at the foot of the previous paragraph. The subsequent steps follow exactly as for Table 11.22*a*. From the statistics set out under the Table we can calculate:

$$b = \frac{180{\cdot}053}{513{\cdot}518} = 0{\cdot}351$$

This value differs from the previous one by little more than 1%; so there is no need to carry the approximations any further.

Table 11.22*b*. *Second approximation by the method of maximum likelihood: Y calculated from the equations* $Y = 2{\cdot}575 + 0{\cdot}347x$, *and* $Y = 3{\cdot}089 + 0{\cdot}347x$

T. confusum

x	n	p	Y	w ($c = 0$)	nw	nwx	y	nwy
3	39	0·08	3·62	0·309	12·051	36·153	3·60	43·384
5	39	0·28	4·31	0·535	20·865	104·325	4·42	92·223
7	39	0·54	5·00	0·637	24·843	173·901	5·10	126·699
9	39	0·69	5·70	0·532	20·748	186·732	5·48	113·699
Total					78·507	501·111 $\bar{x} = 6{\cdot}383$		376·005 $\bar{y} = 4{\cdot}789$

T. castaneum

x	n	p	Y	w ($c = 4$)	nw	nwx	y	nwy
3.5	25	0·17	4·30	0·454	11·350	39·725	4·07	46·195
5·5	25	0·46	5·00	0·588	14·700	80·850	4·90	72·030
7	25	0·79	5·52	0·544	13·600	95·200	5·78	78·608
10	25	0·96	6·56	0·240	6·000	60·000	6·72	40·320
Total					45·650	275·775 $\bar{x} = 6{\cdot}041$		237·153 $\bar{y} = 5{\cdot}195$

	T. confusum	*T. castaneum*	Total
$Snwx^2$	3,527·979	1,850·113	
$\frac{(Snwx)^2}{Snw}$	3,198·597	1,665·977	
S_{xx}	329·382	184·136	513·518
$Snwy^2$	1,833·043	1,266·265	
$\frac{(Snwy)^2}{Snw}$	1,800·855	1,232·016	
S_{yy}	32·188	34·249	66·437
$Snwxy$	2,501·454	1,511·302	
$\frac{Snwx \times Snwy}{Snw}$	2,400·044	1,432·659	
S_{xy}	101·410	78·643	180·053

The lines may now be tested to see whether they differ significantly from parallel. This is done by calculating a series of χ^2 from the sums of squares and products set out under Table 11.22*b*:

$$\chi^2_5 = SS_{yy} - \frac{(SS_{xy})^2}{SS_{xx}} \qquad \text{(Total)}$$

$$= 66{\cdot}437 - \frac{180{\cdot}005^2}{513{\cdot}518}$$

$$= 3{\cdot}31$$

$$\chi^2_2 = S_{yy} - \frac{(S_{xy})^2}{S_{xx}} \qquad (\textit{T. confusum})$$

$$= 32{\cdot}188 - \frac{101{\cdot}40^2}{329{\cdot}382}$$

$$= 0{\cdot}97$$

$$\chi^2_2 = 34{\cdot}249 - \frac{78{\cdot}643^2}{184{\cdot}136} \qquad (\textit{T. castaneum})$$

$$= 0{\cdot}66$$

The sum of the values for the individual species (i.e. 0·97 + 0·66 = 1·63) is subtracted from the total to give a χ^2 with 1 degree of freedom which tests the 'parallelism' of the lines (Table 11.22*c*).

Table 11.22*c*

	Degrees of freedom	χ^2
Departures from parallel	1	1·67
Residual heterogeneity	4	1·63
Total	5	3·31

None of these values for χ^2 is significant. So we may proceed to test the significance of the difference between the estimates of L.D. 50 for the two species without adjusting the data for heterogeneity.

The equations from the second approximation are:

T. confusum, $Y = 2{\cdot}549 + 0{\cdot}351x$
T. castaneum, $Y = 3{\cdot}075 + 0{\cdot}351x$

The value of x that gives $Y = 5$ is the estimated value of L.D. 50. For *T. confusum* L.D. 50 = 6·983 and for *T. castaneum* L.D. 50 = 5·484. The difference is 1·499. The variance of the difference is given by

$$V_d = \frac{1}{b^2}\left[\frac{1}{S_1nw} + \frac{1}{S_2nw} + \frac{(\bar{x}_2 - \bar{x}_1 - d)^2}{SS_{xx}}\right] \quad \text{(See Finney, 1947, equation 5.7.)}$$

$$= \frac{1}{9{\cdot}6221} + \frac{1}{5{\cdot}6241} + \frac{1{\cdot}3387}{63{\cdot}2659}$$
$$= 0{\cdot}3023$$

The standard deviation of the difference is the square root of the variance $= 0{\cdot}5498$, whence we have

$$t = \frac{1{\cdot}499}{0{\cdot}5498}$$
$$= 2{\cdot}726$$

And $P < 0{\cdot}01$. (From Table II of Fisher and Yates, 1948.)

(*f*) Conclusions and comment

We have estimated that about seven days at 5°C is sufficient to kill half the population of *T. confusum* and 5½ days' exposure is sufficient for *T. castaneum*. This is equivalent to saying that, on the average, *T. confusum* can live for about seven days at 5°C and *T. castaneum* for 5½ days.

The null hypothesis that there is no difference between the two species in regard to the time that they can stay alive at 5°C has been disproved with a probability of less than 1%. This is equivalent to saying that the odds against the two species being equally tolerant to exposure to 5°C are rather more than 100 to 1. No matter how often or how elaborately we repeated this experiment we would never be able to assert the inequality of the species with absolute certainty, but only with increasing degrees of probability. This is true of all scientific knowledge.

Compare the results of experiments 12.11 and 11.22. In experiment 12.11 we demonstrated that *T. castaneum* was more strongly influenced by eight days' starvation and desiccation than *T. confusum* (Table 12.11*d*, significant heterogeneity between species after eight days without food) and in experiment 11.22 we have shown that *T. castaneum* is significantly less 'cold-hardy' than *T. confusum*. These facts are consistent with the statement that *T. castaneum* is adapted to living at a higher range of temperature than *T. confusum* and is, on the whole, less 'hardy' than *T. confusum*. This statement, strictly limited though it is to the results of these two experiments, might be regarded as the beginnings of a *theory*. All scientific theory is built up like this by finding generalizations that are consistent with the empirical facts. One can imagine how this theory, even at this early stage in its development, may give rise to speculation and argument leading to hypotheses and experiments.

For example, do other species in this and related genera fit into a graded series according to their 'hardiness' and might this be related to their evolutionary origin – in the tropics or elsewhere. This is the way that scientific knowledge is built up, by proceeding through successive cycles of theory, model, hypothesis, null hypothesis and experiment to theory and so on.

The technique of biological assay and the associated statistical method of probit analysis are part of the essential techniques of ecology. The statistics may seem a little complex at first but with the help of Finney's (1947) useful account of the theory and especially the example that he works out in his Appendix I, the method can be mastered, for practical purposes, fairly easily.

12 *Behaviour in relation to moisture and food*

12.1 MOISTURE

See section 6.21

Experiment 12.11

(*a*) Purpose

To observe the behaviour of the flour beetles *Tribolium confusum* and *T. castaneum* in a gradient of moisture. And to discover (*i*) whether the beetles move into the moist or the dry end of the gradient, (*ii*) whether their behaviour changes after they have been desiccated and starved for a period, and (*iii*) whether the two species differ from each other in their behaviour in a gradient of moisture.

(*b*) Material

About 500 adults of each species.

(*c*) Apparatus

(*i*) 10 specimen tubes (about 25 × 70 mm) with gauze covers.

(*ii*) A desiccator containing a saturated solution of KOH (or concentrated H_2SO_4), and another containing a saturated solution of NaCl or some other solution giving a relative humidity of about 75%.

(*iii*) A constant temperature cabinet set at 27°C.

(*iv*) Several 'humidity-chambers' made as follows. One half of a petri dish (lid or bottom) is placed inside another larger one, and the inner one is packed up, if necessary, until its upper surface is in the same plane as that of the larger dish. A strip of cardboard is used to make a partition that stretches centrally across both dishes and is flush with the top of the dishes as in Fig. 12.11*a*. On one side of the partition the dishes are packed with moist sawdust, on the other side with sawdust that has been oven-dried at 105°C and allowed to cool in a desiccator.

A sheet of nylon gauze is placed over the dishes and then the corresponding half-dishes (if lids have been used for the lower half of the humidity-chamber then lids must also be used for the upper half) are inverted over the gauze. The result, as in Fig. 12.11*a*, is a circular chamber surrounded by a 'guard-ring'. One side of the chamber contains moist air above the moist sawdust and the other dry air. The existence of a gradient may be demonstrated by placing a strip of cobalt-thiocyanate paper across the chamber at right angles to the partition. The guard-ring reduces the chances of secondary gradients developing around the

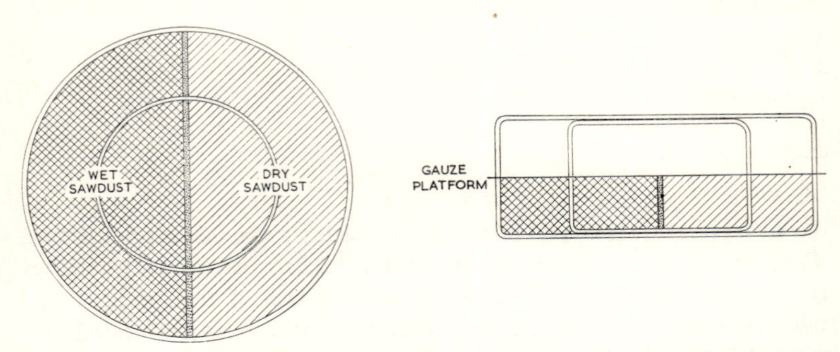

Fig. 12.11*a*. Humidity-chamber for measuring the response of *Tribolium* to a gradient in moisture. The beetles are placed in the inner compartment. The guard-ring serves to reduce the influence of secondary gradients which develop at the margin. The central partition comes flush with the top of the lower dish. The platform of nylon gauze on which the beetles move separates the lower dishes from the upper.

edges. To use the humidity-chambers the lids are lifted briefly, the beetles scattered over the gauze platform and the lids replaced immediately.

(*d*) Method

(*i*) Eight days before the experiment is to be done, collect 500 adults of each species. Divide each lot at random into 10 samples of 50 each. Place two lots of 50 into clean, dry tubes and put them in the desiccator at 15% relative humidity (saturated solution of KOH) at 27°C. Place the remaining eight samples in tubes with a little flour-medium and put them at 70% relative humidity at 27°C.

(*ii*) At two-day intervals sieve out two lots of beetles from the flour-medium and place them in clean, dry tubes at 15% relative humidity. By the eighth day there will be two lots of 50 beetles that have been kept in the desiccator at 15% R.H. and 27°C without food for 2, 4, 6 and 8 days and two lots of 50 that have been in moist flour-medium all the time.

(*iii*) The next step is to set up the humidity-chambers and allow them to stand for 15 to 20 minutes to settle down. Then each sample of 50 beetles is tipped into a humidity-chamber and their behaviour watched and described. At intervals of one minute for upwards of 15 minutes the number on either side of the partition should be counted and recorded. The beetles respond best at temperatures above 25°C. If the temperature of the laboratory is below 25°C it may be necessary to do the experiment in a constant temperature room.

(*e*) Results

The result of an experiment in which *Tribolium castaneum* were kept without food for 0, 4 and 8 days before being tested in the humidity-chamber are set out in Table 12.11*a* and Fig. 12.11*b*. The replicates agree fairly well so Fig. 12.11*b* was drawn by combining the two replicates and transforming the totals to percentages.

From Table 12.11*a* we may interpolate to find the number of beetles on the dry side after 10 minutes, which may be taken as a measure of

Table 12.11*a*. *The number of Tribolium castaneum on the dry side of the humidity chamber*

Minutes after release	Days without food					
	0 days		4 days		8 days	
	First replicate	Second replicate	First replicate	Second replicate	First replicate	Second replicate
1	20	33	15	18	6	1
3	38	36	22	20	6	3
5	38	35	20	12	2	2
7	34	30	22	20	5	2
9	35	35	20	23	2	3
11	37	33	21	22	3	3
13	38	29	20	19	3	3
15	30	33	15	18	4	1
20	30		20		5	
No. in replicate	50	50	50	50	21	23

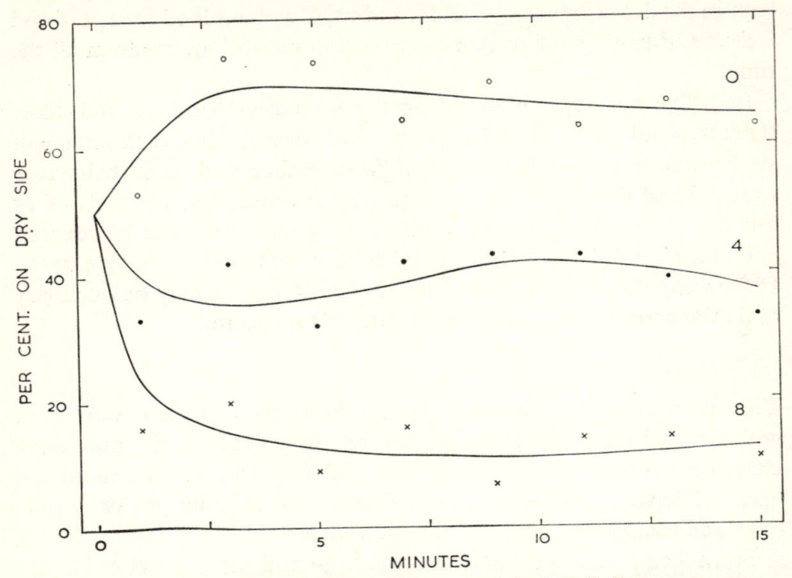

Fig. 12.11*b*. The number (expressed as a per cent) of *Tribolium castaneum* on the dry side of the humidity-chamber. The figures alongside the curves indicate the number of days of starvation before the experiment.

the beetles' response to the gradient of humidity. A similar experiment was done with *T. confusum*. The results for both species are shown in Table 12.11*b*.

About 70% of the beetles that had been living in the normal flour-medium up to the time of the experiment accumulated on the dry side of the humidity-chamber. But a majority of the beetles that had been without food for eight days (87% of *T. castaneum* and 68% of *T.*

Table 12.11*b*. *The behaviour of Tribolium confusum and T. castaneum in a gradient of moisture*

Days without food	*T. confusum*			*T. castaneum*		
	No. on dry side	Total	% on dry side	No. on dry side	Total	% on dry side
0	74	98	76	70	100	70
4	42	100	42	43	100	43
8	24	78	31	6	44	13

confusum) were found on the wet side. These figures suggest (*i*) that the beetles recognize a gradient in moisture and tend to accumulate in one end or the other of it. (*ii*) That their response to the gradient depends on their condition – the normal tendency to accumulate on the dry side being reversed after eight days' starvation. (*iii*) That this reversal of behaviour was more pronounced in *T. castaneum* than in *T. confusum*.

To measure the reliability of these conclusions we have to disprove the three null hypotheses (*i*) that the number of beetles on the dry side was not significantly different from the number on the wet side. (*ii*) That the distribution of the beetles did not depend on the number of days that they had been without food. (*iii*) That the distribution of

Table 12.11c

Species	Days without food	No. of beetles	No. on the dry side		χ^2 $\frac{2(O-E)^2}{E}$	P
			Observed	Expected		
T. confusum	0	98	74	49	25·5	<0·001
	4	100	42	50	2·6	N.S.
	8	78	24	39	11·5	<0·001
T. castaneum	0	100	70	50	16·0	<0·001
	4	100	43	50	2·0	N.S.
	8	44	6	22	23·3	<0·001
Total		520	259	260	0·004	N.S.

Total $\chi^2_6 = 80{\cdot}9$, $P < 0{\cdot}001$

T. confusum was not significantly different from the distribution of *T. castaneum*.

If the beetles do not recognize a gradient in humidity there should, on the average, be equal numbers of them on either side of the humidity-chamber and any discrepancy from the 1 : 1 ratio would have to be explained as errors of random sampling. By calculating a series of χ^2 as in Table 12.11*c* we can estimate the odds against the observed discrepancies having arisen by chance. The sum of the six values of χ^2 in Table 12.11*c* is a χ^2 with six degrees of freedom. It is highly significant; there is less than one chance in a thousand that the observed discrepancies from the 1 : 1 ratio arose as errors of random sampling; apparently the beetles do recognize a gradient in moisture and distribute themselves unevenly in response to it. The difference between 80·9 and 0·004 gives a χ^2 with five degrees of freedom which measures the significance of the heterogeneity between the individual treatments. The heterogeneity is highly significant ($P < 0{\cdot}001$). An inspection of the Table shows clearly that by far the greatest part of this heterogeneity is associated with the reversal of the behaviour of the beetles after they have been without food. Four of the six values for χ^2 are individually significant. The departures from the 1 : 1 ratio are significant for all the tests in the zero and eight day series. The fact that the results for zero days are in the opposite direction from those for eight days and that both are significantly different from the 1 : 1 ratio disproves ($P < 0\ 001$) the second null hypothesis and establishes the conclusion that the response of the beetles to the gradient in moisture depends on the number of days that they have been without food.

But one wonders whether the difference between the species at eight days (i.e. the relatively greater reversal in the behaviour of *T. castaneum*) may also contribute significantly to the heterogeneity between treatments. Because the analysis in Table 12.11*c* has established a significant discrepancy from the 1 : 1 ratio and a significant heterogeneity between days, the test for homogeneity between species must not assume either a 1 : 1 ratio or homogeneity between days. The correct procedure is to construct a 2×2 contingency table for each day separately as in Table 12.11*d*.

There was significant heterogeneity between the species at eight days ($\chi^2 = 4{\cdot}4$) but not at 0 or 4 days. The heterogeneity at eight days was due to the relatively greater response by *T. castaneum* to the period of starvation (Table 12.11*b*).

Table 12.11*d*. *Test for homogeneity between species*

Days without food		*T. confusum*		*T. castaneum*		Total obs.	χ^2 $\frac{S(O-E)^2}{E}$	P
		Obs.	Exp.	Obs.	Exp.			
0	Dry side	74	71·3	70	72·7	144	0·74	N.S.
	Wet side	24	26·7	30	27·3	54		
	Total	98	98	100	100	198		
4	Dry side	42	42·5	43	42·5	85	0·02	N.S.
	Wet side	58	57·5	57	57·5	115		
	Total	100	100	100	100	200		
8	Dry side	24	19·2	6	10·8	30	4·42	<0·05
	Wet side	54	58·8	38	33·2	92		
	Total	78	78	44	44	122		

(*f*) Conclusions and comment

Beetles of both species in their normal freshly fed condition moved towards the dry side of the gradient and tended to stay there. Beetles that had been starved for eight days moved towards the wet side of the gradient. There was no significant difference in the behaviour of the two species in the 0- and 4-day lots. About half the beetles that had been without food for four days were found on either side of the humidity-chamber. Perhaps half the beetles retained their original preference for dry air while the other half had come to prefer moist air. Or, perhaps all the beetles lost their capacity to recognize a gradient in moisture. This experiment does not allow us to choose between these two explanations. The difference between the species after eight days without food suggests that *T. confusum* may be 'hardier' than *T. castaneum*.

The results of this experiment open up a number of questions that can be explored with a series of experiments, using the same techniques. It is a good opportunity for students to exercise their own initiative in proposing and designing experiments. I mention three matters that are obviously worth investigating, but there are others.

(*a*) In this experiment the beetles were both starved and desiccated. How can the relative importance of these two treatments be sorted out?

(*b*) In this experiment no notice was taken of gradients in other than moisture. Light is an obvious one that might have been confounded in the results.

(*c*) In this experiment no effort has been made to explain the results with the intermediate treatments where the distribution of the beetles relative to the gradient in moisture was not significantly different from a random scatter. Brush up your philosophy of science before you follow this one too far.

12.2 FOOD

Experiment 12.21

(*a*) Purpose

To discover whether the females of the white cabbage butterfly *Pieris rapae* are more likely to lay their eggs on Brassicae than on other sorts of plants that occur commonly in gardens where Brassicae may be grown.

(*b*) Material

About 30 adults of *Pieris rapae*. We usually rely on catching them in the field rather than take the trouble to maintain a laboratory stock.

(*c*) Apparatus

(*i*) About 30 six-inch flower-pots and suitable potting soil.

(*ii*) Six seedlings of cabbage or some other Brassicae, *Nasturtium*, *Geranium*, *Sonchus* and lettuce, or any other similar group of plants. Also some seeds or seedlings of *Alyssum*, *Erigeron* or some other sort of plant that will grow well in pots and provide plenty of flowers with nectar for the butterflies to feed on.

(*iii*) A cage of fly-wire gauze about 2 × 1 × 1 m. A convenient cage has a trap-door lid through which the pots may be arranged, and cloth sleeves in the sides which allow access to the cage after the butterflies have been put in.

(*d*) Method

(*i*) Some weeks before the experiment is to be done the plants should be established in the pots. It may take longer to get the flowering plants going well.

(*ii*) When the experiment is to be done place several pots of flowering plants at one end of the cage and arrange the plants for egg-laying at the other end. Use five of each sort and arrange the 25 pots in the form of a randomized block. In this instance the randomized block has five rows with five plants in each row. Each row contains one and only one of each sort of plant. And within each row the positions of the different sorts of plants occur at random. This arrangement is best attained by numbering the pots and then arranging them in the rows according to numbers drawn from a table of random numbers. The arrangement that was used in the present experiment is shown in Table 12.21*a*.

Table 12.21*a*

						Key
I	A	C	E	D	B	A *Nasturtium*
II	E	D	A	B	C	B *Geranium*
III	A	C	B	D	E	C Brassicae
IV	E	B	A	D	C	D Lettuce
V	A	B	D	E	C	E *Sonchus*
				Alyssum		

(*iii*) After the plants have been arranged, catch about 30 butterflies and let them go in the cage. Leave them there for several days. This experiment is best done in the open air or in a glass house in sunny weather, because the butterflies lay eggs more freely when the sun is shining.

(*iv*) After several days take the plants out of the cage and count the eggs that have been laid on them. Take care not to dislodge the eggs. Keep a record of the position of each plant and the number of eggs on it as in Table 12.21*b*.

(*e*) Results

The results of a typical experiment are shown in Table 12.21*b*.

The differences in the mean number of eggs laid on the Brassicae and on all the other sorts of plants are so great and so consistent that we may take them as highly significant without troubling to calculate precise probabilities. The extreme heterogeneity in the variances would, in any case, make the calculation of precise odds impracticable. The

Table 12.21*b*. *The number of eggs laid by Pieris rapae on five different sorts of plants, arranged in a randomized block as indicated in Table 12.21a*

						Total		
I	12	42	0	0	0	54	*Nasturtium*	51
II	0	0	12	3	24	39	*Geranium*	4
III	6	81	1	0	0	88	Brassicae	359
IV	0	0	13	0	105	118	Lettuce	0
V	8	0	0	0	107	115	*Sonchus*	0
					Total	414	Total	414

difference between the number of eggs on *Nasturtium* and *Geranium* also seems to be quite beyond doubt but in this case one may, if one wishes, calculate the precise significance of the difference between the means because the variances seem to be reasonably homogeneous. The variance of the difference between the means is given by:

$$s^2 = \frac{1}{n_1 + n_2 - 2}[S(x_1 - \bar{x}_1)^2 + S(x_2 - \bar{x}_2)^2]$$

For the significance of the difference between means we require to estimate

$$t = \frac{\bar{x} - \bar{x}_2}{s} \times \sqrt{\left(\frac{n_1 \times n_2}{n_1 + n_2}\right)}$$

In the present instance:

$$s^2 = \tfrac{1}{8}(36{\cdot}8 + 6{\cdot}8)$$
$$s = 2{\cdot}334$$
$$t = \frac{10{\cdot}2 - 0{\cdot}8}{2{\cdot}334} = 6{\cdot}4$$
$$P < 0{\cdot}001$$

(*f*) Conclusion and comment

The tendency for *Pieris* to lay most of its eggs on Brassicae in preference to *Nasturtium*, *Geranium*, *Sonchus* or lettuce has been established at a very high though unknown level of probability. The tendency to lay more eggs on *Nasturtium* than on *Geranium* was established with a probability of less than 0·001.

The heterogeneity of the variances over the whole experiment precluded a full analysis of the randomized block which would have given an estimate of the significance of the difference between rows. But there is just a suggestion that the rows near the *Alyssum* are likely to have more eggs laid on them than the others.

It so happened that, in this experiment, the five Brassicae comprised one each of cabbage, broccoli, brussels sprout, cauliflower and greenleaf cauliflower. The number of eggs laid was greatest (107) on the broccoli and least (24) on the brussels sprout. Was this just an accident of random sampling or does it reflect a tendency that is likely to be repeated on another occasion? The present experiment gives no answer to these questions because the varieties of Brassicae were not replicated. So we design experiment 12.22.

Experiment 12.22

(*a*) Purpose

To discover whether *Pieris rapae* are likely to lay different numbers of eggs on different varieties of Brassicae when they are given equal opportunity to lay on any of them.

(*b*) Material

About 30 adult *Pieris rapae*.

(*c*) Apparatus

The same as for 12.21 except that the 25 plants for egg-laying comprise five varieties of Brassicae, e.g. broccoli, cabbage, brussels sprout, cauliflower and greenleaf cauliflower.

(*d*) Method

The same as for experiment 12.21 except that the plants are arranged in a latin square instead of a randomized block. In the latin square each variety is represented once and once only in each row and once and once only in each column. Apart from this restriction their positions in the rows and columns are random. This design allows gradients, if they exist, to be picked up across rows and across columns, and in addition enhances the precision of the comparison of varieties. In the present instance the comparisons between rows and columns are desirable because it seems likely that the presence of the *Alyssum* at one end of the cage may cause a gradient in the number of eggs laid; also vagaries of sun, shadow, wind, etc. may cause a gradient in another direction.

The method for selecting a latin square at random is given by Fisher and Yates (1948), Table XV. The design that was used in the present experiment is shown in Table 12.22*a*.

Table 12.22*a*

	Design of latin square (i)	(ii)	(iii)	(iv)	(v)	Key
I	B	D	E	A	C	A Greenleaf cauliflower
II	C	E	A	B	D	B Cauliflower
III	D	C	B	E	A	C Broccoli
IV	A	B	C	D	E	D Cabbage
V	E	A	D	C	B	E Brussels sprout
				Alyssum		

Table 12.22*b*. *Number of eggs laid by Pieris on five varieties of Brassicae arranged in a latin square*

	(i)	(ii)	(iii)	(iv)	(v)	Total	Varieties	Eggs
I	42	18	30	6	20	116	Greenleaf cauliflower	110
II	54	4	3	6	18	85	Cauliflower	83
III	4	5	2	12	53	76	Broccoli	122
IV	15	16	9	14	33	87	Cabbage	70
V	13	33	16	34	17	113	Brussels sprout	92
Total	128	76	60	72	141	477	Total	477

(*e*) Results

The number of eggs on each plant are given in Table 12.22*b* in the positions in which they occurred in the experiment.

From an inspection of the Table it may be seen that more eggs were laid on broccoli than on any other variety and also more eggs were laid round the edges than in the centre of the square. The differences are only moderate and it is rather doubtful whether they will be significantly greater than the variances within varieties. An analysis of variance will answer this question precisely. For the analysis of variance the sums of squares for total, rows, columns and varieties are calculated as follows,

$$\text{Sum of squares,} \qquad S(x - \bar{x})^2 = \frac{Sx^2}{n'} - \text{C.F.}$$

For rows, x is the total for each individual row

$$n' = 5$$

For columns, x is the total for each individual column

$$n' = 5$$

For varieties, x is the total for each individual variety

$$n' = 5$$

For total, x is each individual figure in the body of the Table

$$n' = 1$$

The correction factor C.F. is the same for all expressions

$$\text{C.F.} = \frac{(Sx)^2}{n} = \frac{477^2}{25}$$

The sums of squares are set out in Table 12.22*c*.

Table 12.22*c*

	$\frac{Sx^2}{n'}$	C.F.	$S(x - \bar{x})^2$
Rows	9,359	9,101·16	257·84
Columns	10,165	9,101·16	1,063·84
Varieties	9,447·4	9,101·16	346·24
Total	14,529	9,101·16	5,427·84

The analysis of variance is given in Table 12.22*d*.

Table 12.22*d* *Analysis of variance*

Source of variance	Degrees of freedom	Sums of squares	Mean square	Variance ratio	*P*
Columns	4	1,063·84	265·96	<1	N.S.
Rows	4	257·84	64·46	<1	N.S.
Varieties	4	346·24	86·56	<1	N.S.
Residual	12	3,759·92	313·33		
Total	24	5,427·84			

None of the variance ratios is significant.

(*f*) Conclusion and comment

There is no evidence from this experiment that the distribution of the eggs depended on the position of the row or the column or on variety. This is not to say that significant differences might not be demonstrable with a larger experiment. With a sufficiently large number of replicates even quite small differences may be shown to be significant.

Experiment 12.23

A population of snails *Helicella virgata* living in open parkland, when measured by the methods of section 10.3, was found to be patchily distributed. The snails seemed to be concentrated in places where the normally sparse cover of low grasses and forbs was less sparse than usual. During wet weather the snails were often seen crawling over the surface of the ground and feeding. From an examination of their faeces it seemed that the snails fed mostly by scraping the surface of the ground. When it was not actually raining, but the ground was wet, the snails spent most of their time resting on grass-stems and other vegetation at least an inch or two above the surface of the ground. Snails that were artificially confined continuously on wet soil became bloated with water and unhealthy. One hypothesis to explain the patchy distribution of the snails is that they congregate where there is shelter and food. Shelter is in the form of grass-stems to climb on and so avoid the risk of resting too long on wet ground; food in the form of decayed or decaying organic matter to be scraped from the surface of the soil.

(*a*) Purpose

To test the hypothesis that the snails *Helicella virgata* congregate in places where they can find adequate shelter and food.

(*b*) Materials

A natural population of *H. virgata*.

(*c*) *Apparatus*

A point-quadrat which is a metal bar mounted on legs. The bar is perforated by a number of vertical holes which serve as close guides for thin metal rods which slide vertically through these guides. A point-quadrat can be used for estimating the density of a sward of pasture or other low vegetation by counting the number of leaves or stems touched by the steel rods as they slide down towards the ground. A gardener's trowel or other implement for scraping up the surface layer of soil. Containers for holding the samples to carry them back to the laboratory. And large beakers or other vessels that will serve to separate the organic matter from the rest of the soil by flotation in water.

(*d*) Methods

Choose about 20 quadrats of a suitable size, say about one square foot. Choose quadrats in different sorts of places so that the variability in the snail numbers, the vegetation and the organic matter is adequately represented in the sample. Count the snails; estimate the density of the vegetative cover using the point-quadrat; and collect the top 0·5 cm or so of soil. Separate the organic matter from the rest of the soil by floating it on water. Discard any living leaves or stems that may be in the sample because this is not food for *H. virgata*. Dry the organic matter and weigh it.

(*e*) Results

In one experiment we counted 19 quadrats, and got the results shown in Table 12.23*a*. The shelter is recorded as the number of 'contacts' made by the rods of the point-quadrat; the organic matter in grams of air-dry material. The extent to which the number of snails on a quadrat is associated with the vegetation and the organic matter can be measured by estimating the constants in the equation:

$$Y = \bar{y} + b_1(x_1 - \bar{x}_1) + b_2(x_2 - \bar{x}_2)$$

The constant b_1 measures the linear regression of y on x_1 independent of the association of y_1 with x_2, and allowing for any association of x_1 with x_2. Similarly b_2 measures the linear regression of y on x_2 independent of its association with x_1.

Table 12.23*a*

	Quadrat	Number of snails		Vegetation contacts	Organic matter g
		Predicted Y	Observed y	x_1	x_2
	1	14·05	9	11	3·3
	2	19·83	18	3	11·5
	3	13·62	17	5	6·2
	4	11·52	16	7	3·7
	5	17·91	10	12	5·4
	6	15·62	13	10	4·9
	7	5·42	4	2	2·2
	8	6·54	7	4	1·9
	9	4·44	4	3	1·0
	10	6·69	3	1	3·6
	11	5·13	7	2	2·0
	12	20·80	16	8	9·5
	13	7·61	6	5	2·1
	14	23·49	21	9	10·8
	15	26·06	28	17	8·3
	16	21·88	27	9	9·7
	17	26·01	28	19	7·2
	18	13·57	26	7	5·1
	19	7·81	8	6	1·7
Sx	19	268·00	268	140·0	100·1
Sx^2	—		5,128·00	1,468·00	729·87
$\frac{(Sx)^2}{n}$	—		3,780·21	1,031·57	527·36
$S(x-\bar{x})^2$	—		1,347·79	436·43	202·51
Syx_1	2,493·00	Syx_2	1,803·40	Sx_1x_2	858·90
$\frac{Sy \times Sx_1}{n}$	1974·73	$\frac{Sy \times Sx_2}{n}$	1411·93	$\frac{Sx_1 \times Sx_2}{n}$	737·57
$S(y-\bar{y})(x_1-\bar{x}_1)$	518·27	$S(y-\bar{y})(x_2-\bar{x}_2)$	391·47	$S(x_1-\bar{x}_1)(x_2-\bar{x}_2)$	121·33

$$b_1 = \frac{S(x_2-\bar{x}_2)^2 \times S(x_1-\bar{x}_1)(y-\bar{y}) - S(x_1-\bar{x}_1)(x_2-\bar{x}_2) \times S(x_2-\bar{x}_2)(y-\bar{y})}{S(x_1-\bar{x}_1)^2 \times S(x_2-\bar{x})^2 - [S(x_1-\bar{x}_1)(x_2-\bar{x}_2)]^2}$$

$$b_2 = \frac{S(x_1-\bar{x}_1)^2 \times S(x_2-\bar{x}_2)(y-\bar{y}) - S(x_1-\bar{x}_1)(x_2-\bar{x}_2) \times S(x_1-\bar{x}_1)(y-\bar{y})}{S(x_1-\bar{x}_1)^2 \times S(x_2-\bar{x}_2)^2 - [S(x_1-\bar{x}_1)(x_2-\bar{x}_2)]^2}$$

The quantities required for the estimation of b_1, b_2 are given at the top of page 258.

$$b_1 = \frac{202{\cdot}51 \times 518{\cdot}27 - 121{\cdot}33 \times 391{\cdot}47}{436{\cdot}43 \times 202{\cdot}51 - 121.33^2}$$
$$= 0{\cdot}7800$$
$$b_2 = \frac{436{\cdot}43 \times 391{\cdot}47 - 121{\cdot}33 \times 518{\cdot}27}{436{\cdot}43 \times 202{\cdot}51 - 121.33^2}$$
$$= 1{\cdot}4657$$

So the regression equation reads:

$$Y = a + b_1x_1 + b_2x_2$$
$$= 0{\cdot}636 + 0{\cdot}780\, x_1 + 1{\cdot}466\, x_2$$

where $a = \bar{y} - b_1\bar{x}_1 - b_2\bar{x}_2$

The values of Y for the 19 quadrats got by substituting the values for x_1 and x_2 are given in column two of Table 12.23*a*. The agreement of SY with Sy is a cross-check for the accuracy of the foregoing arithmetic. The significance of the regression is estimated in Table 12.23*b*.

Table 12.23*b*

Source of variance	Degrees of freedom	Sums of squares	Mean squares	Variance ratio	P
Total, $S(y - \bar{y})^2$	18	1,347·79	—	—	—
b_1	1	404·25	404·25	17·49	$<0{\cdot}001$
b_2	1	573·78	573·78	24·82	$<0{\cdot}001$
Regression $S(Y - \bar{y})^2$	2	978·03	489·02	21·16	$<0{\cdot}001$
Residual	16	369·76	23·11	—	—

The sums of squares for regression in column three are estimated as follows:

Sum of squares for b_1 $= b_1 \times S(x_1 - \bar{x}_1)(y - \bar{y})$
$= 0{\cdot}7800 \times 518{\cdot}27$
$= 404{\cdot}251$

Sum of squares for b_2 $= b_2 \times S(x_2 - \bar{x}_2)(y - \bar{y})$
$= 1{\cdot}4657 \times 391{\cdot}47$
$= 573{\cdot}78$

Sum of squares for regression $= 404{\cdot}251 + 573{\cdot}78$
$= 978{\cdot}03$

Residual sum of squares = Total S.S. − Regression S.S.
$= 1347{\cdot}79 - 978{\cdot}03$
$= 369{\cdot}76$

The percentage of the total variance to be attributed to the regression is estimated:

$$R^2 = \frac{S(Y - \bar{y})^2}{S(y - \bar{y})^2} = \frac{978{\cdot}03}{1347{\cdot}79} = 0{\cdot}7256$$

$$1 - A = \frac{18(1 - R^2)}{16} = 0{\cdot}6913$$

$$\text{Percent variance} = 100\,A$$
$$= 69\%$$

To estimate the standard error of the regression coefficients first estimate the quantities C_{11} and C_{22}.

$$C_{11} = \frac{S(x_2 - \bar{x}_2)^2}{S(x_1 - \bar{x}_1)^2 \times S(x_2 - \bar{x}_2)^2 - [S(x_1 - \bar{x}_1)(x_2 - \bar{x}_2)]^2}$$

$$C_{22} = \frac{S(x_1 - \bar{x}_1)^2}{S(x_1 - \bar{x}_1)^2 \times S(x_2 - \bar{x}_2)^2 - [S(x_1 - \bar{x}_1)(x_2 - \bar{x}_2)]^2}$$

$$\text{S.E. of } b_1 = \sqrt{(s^2 \times C_{11})}$$

$$\text{S.E. of } b_2 = \sqrt{(s^2 \times C_{22})}$$

where s^2 is the residual mean square from Table 12.23*b*.

$$\text{So for } b_1,\ t_{16} = \frac{b_1}{\sqrt{(s^2 \times C_{11})}}$$
$$= 3{\cdot}09;\ P < 0{\cdot}01$$

$$\text{for } b_2,\ t_{16} = \frac{b_2}{\sqrt{(s^2 \times C_{22})}}$$
$$= 3{\cdot}96;\ P < 0{\cdot}01$$

In order to estimate the relative importance of b_1 and b_2 the units in which the regression coefficients are measured must be transformed to standard measure because the original units (contacts with a point-quadrat and grams) are not additive. The regression coefficients in units of standard measure are given by:

$$b_1^1 = b_1\sqrt{\left(\frac{S(x_1 - \bar{x}_1)^2}{S(y - \bar{y})^2}\right)}$$
$$= 0{\cdot}78\sqrt{\left(\frac{436}{1348}\right)}$$
$$= 0{\cdot}44$$

$$b_2^1 = b_2\sqrt{\left(\frac{S(x_2 - \bar{x}_2)^2}{S(y - \bar{y})^2}\right)}$$
$$= 1{\cdot}466\sqrt{\left(\frac{202}{1348}\right)}$$
$$= 0{\cdot}57$$

(*f*) Conclusion and comment

The null hypothesis has been disproved at a high level of probability ($P < 0{\cdot}001$) for the regression as a whole and for each coefficient individually. The correlation between x_1 and x_2 was small ($r = 0{\cdot}41$) and barely significant ($P = 0{\cdot}04$) which may be why both regression coefficients proved to be individually significant. The density of the vegetation was not a good indication of the amount of organic matter but the snails responded to both. In standard measure b_2^1 was larger than b_1^1 which suggested that the snails responded more strongly to organic matter (food) than to vegetation (shelter).

Having disproved the null hypothesis we know that the facts, as we have measured them, are consistent with the explanation that we have chosen to regard as most plausible namely, that the snails congregate where there is shelter and food. This explanation may stand until someone thinks of a more plausible explanation that is not only consistent with these facts but also with new facts that our theory fails to explain. This final step in the scientific process is inevitably subjective (see also section 8.2).

Appendix

MATERIALS FOR EXPERIMENTS

The following species have been used in the experiments described in Part II.

Paramoecium sp., *Helix aspersa*, *Helicella virgata*, *Thrips imaginis*, *Saissetia oleae*, *Calandra oryzae*, *Calandra granaria*, *Tribolium castaneum* *Tribolium confusum*, *Tenebrio molitor*, *Pieris rapae*.

We do not keep laboratory colonies of *Helix*, *Helicella*, *Thrips* *Saissetia*, or *Pieris* because we find it easier to catch wild ones in the field when we need them. Laboratory colonies of the other species may be maintained as follows.

Paramoecium

Glass jars containing about one litre are suitable. The medium is made by boiling lawn-cuttings for about 30 minutes in tap-water, allowing it to cool, straining it and then leaving it on the bench for five days until it is heavily populated by bacteria. It is then inoculated with a few *Paramoecium*. At room temperatures the colony usually becomes quite dense in about five days and after about seven days it requires to be subcultured.

Calandra oryzae and *C. granaria*

Tin cans with press-in lids and a capacity of about two litres are quite good. Ventilators of fine brass gauze (about 10 mesh to the centimetre) may be soldered in the lid and in the bottom. Soft wheat of about 11% water-content is suitable for food. The can is about two-thirds filled with wheat; a hundred or so adults, or a handful of wheat from an established colony are added and the can is stored at 27°C 70% R.H. The stock may be maintained indefinitely if it is sub-cultured about every two months. It is convenient to accumulate four or five colonies and then, as each new one is started, the oldest is discarded.

Tribolium confusum and *T. castaneum*

Glass jars with a capacity of 1 to 1·5 l. are quite good. There is no need for covers if they are not filled beyond about three-quarters. The medium is made by mixing about 10% powdered yeast with white flour. Temperature and humidity are not critical but the beetles thrive best at a relative humidity of about 60% and at temperatures between 27° and 30°C. Begin by putting 100 adults into about 200 g of medium. After a week sieve out the adults and place them in a fresh jar. Return the egg-laden flour to the same jar and add about 100 g of medium. Repeat this each week until adults begin to appear in the first jar. Then as each new colony is founded throw the oldest one away. At 27°C this usually results in the maintenance of six or eight colonies.

Tenebrio molitor

Wooden boxes with sliding lids, measuring about 40 × 40 × 20 cm, are suitable containers for rearing *Tenebrio*. The medium is a mixture, in equal parts, of bran, pollard and wheat germ. Place about 10 layers of jute sackcloth in the box and place several handfuls of medium between each layer. Add 100 adults and place the box at about 27°C. If a few pieces of fresh carrot or banana are kept in the box for the adults they are likely to lay more eggs. At 27°C the life-cycle occupies several months and at lower temperatures much longer. So it is desirable to maintain a number of colonies in different stages.

Bibliography and author index

(Numbers in italics denote references to text)

ALLEE, W. C., EMERSON, A. E., PARK, O., PARK, T. and SCHMIDT, K. P. (1949) *Principles of Animal Ecology*, Philadelphia: W. B. Saunders & Co. *3, 6, 20, 100*

ANDREWARTHA, H. G. (1943) 'The significance of grasshoppers in some aspects of soil conservation in South Australia and Western Australia', *J. Dept. Agric. South Australia* **46,** 314–22. *49*

(1944) 'The distribution of plagues of *Austroicetes cruciata* Sauss (Acrididae) in Australia in relation to climate, vegetation and soil', *Trans. Roy. Soc. South Australia,* **68,** 315–26. *8, 49, 82*

(1952) 'Diapause in relation to the ecology of insects', *Biol. Rev.* **27,** 50–107. *93*

(1959) 'Self-regulatory mechanisms in animal populations', *Australian J. Sci.* **22,** 200–5. *83*

ANDREWARTHA, H. G. and BIRCH, L. C. (1954) *The Distribution and Abundance of Animals,* Chicago: University of Chicago Press. *8, 19, 25, 54, 76, 87, 129, 136, 144, 156*

ANDREWARTHA, H. G. and BROWNING, T. O. (1961) 'An analysis of the idea of "resources" in animal ecology', *J. theoret. Biol.* **1,** 83–97. *25, 31, 33, 35, 50, 129*

ANDREWARTHA, H. G., MONRO, J. and RICHARDSON, N. L. (1967) 'The use of sterile males to control populations of Queensland fruit-fly *Dacus tryoni* (Frogg.) Diptera Tephritidae', *Australian J. Zool.* **15,** 475–99. *57*

AUMANN, G. D. and EMLEN, J. T. (1965) 'Relation of population density to sodium availability and sodium selection by microtine rodents', *Nature, Lond.* **208,** 198. *82*

BAILEY, N. T. J. (1951) 'On estimating the size of mobile populations from recapture data', *Biometrika* **38,** 293–306. *178*

(1952) 'Improvements in the interpretation of recapture data', *J. Anim. Ecol.* **21,** 120–7. *178*

BAKER, J. R. and RANSOM, R. M. (1932) 'Factors affecting the breeding of the field mouse (*Microtus agrestis*). I. Light', *Proc. R. Soc. Lond., B* **110,** 313–21. *127*

BANKS, C. J. (1954) 'A method of estimating populations and counting large numbers of *Aphis fabae* Scop.', *Bull. Entom. Res.* **45,** 751–6. *166*

BARBER, G. W. (1939) 'Injury to sweet corn by *Euxesta stigmatias* Loew in Florida', *J. Econ. Ent.* **32,** 879–80. *183*

BARTON-BROWNE, L. (1958) 'The choice of communal oviposition sites by the Australian sheep blowfly *Lucilia cuprina*', *Australian J. Zool.* **6,** 241–7. *135*

BATEMAN, M. A. (1955) 'The effect of light and temperature on the rhythm of pupal ecdysis in the Queensland fruit-fly *Dacus* (*Strumeta*) *tryoni* (Frogg.)', *Australian J. Zool.* **3,** 22–33. *124*

BATEMAN, M. A., FRIEND, A. H. and HAMPSHIRE, F. (1966) 'Population suppression in the Queensland fruit-fly *Dacus* (*Strumeta*) *tryoni* (Frogg.). II. Experiments on isolated populations in western New South Wales', *Australian J. Agric. Res.* **17,** 699–718. *57*

BEALL, G. (1946) 'Seasonal variation in sex proportion and wing-length in the migrant butterfly, *Danaus plexippus* L. (Lep. Danaidae)', *Trans. Roy. Ent. Soc., Lond.* **97,** 337–53. *185*

BEARD, R. L. (1945) 'Studies on the milky disease of Japanese beetle larvae', *Conn. Agr. Exp. Sta. Bull.* **491,** 505–81. *73*

BILLINGS, F. H. and GLENN, P. A. (1911) 'Results of the artificial use of white-fungus disease in Kansas', *U.S.D.A. Bur. Entom. Bull.* **107.** *73*

BIRCH, L. C. (1946) 'The heating of wheat stored in bulk in Australia', *J. Australian Inst. Agric. Sci.* **12,** 27–31. *89*

BIRCH, L. C. and ANDREWARTHA, H. G. (1941) 'The influence of weather on grasshopper plagues in South Australia', *J. Dept. Agric. South Australia* **45,** 95–100. *134*

BIRD, F. T. (1953) 'The use of a virus disease in the biological control of the European pine sawfly', *Canad. Entomol.* **85,** 437–46. *73*

(1955) 'Virus diseases of sawflies', *Canad. Entomol.* **87,** 124–7. *73*

BISSONNETTE, T. H. (1935) 'Modification of mammalian sex cycles. III. Reversal of the cycle in male ferrets (*Putorius vulgaris*) by increasing periods of exposure to light between October second and March thirtieth', *J. exp. Zool.* **71,** 341–67. *128*

BOGUSLAVSKY, G. W. (1956) 'Statistical estimation of the size of a small population', *Science, N.Y.* **124,** 317–18. *177*

BRAIDWOOD, R. J. and REED, C. A. (1957) 'The achievement and early consequences of food-production: a consideration of the archeological and natural-history evidence', *Symp. quant. Biol.* **22,** 19–31. *29*

BRETT, J. R. (1944) 'Some lethal temperature relations of Algonquin Park fishes', *Univ. Toronto Stud. Biol.* **52,** 1–49. *96*

BROWNING, T. O. (1952) 'The influence of temperature on the rate of development of insects, with special reference to the eggs of *Gryllulus commodus* Walker', *Australian J. scient. Res. B* **5,** 96–111. *229*

(1954) 'Water balance in the tick *Ornithodorus moubata* Murray, with particular reference to the influence of carbon dioxide on the uptake and loss of water', *J. exp. Biol.* **31,** 331–40. *114*

(1959) 'The long-tailed mealybug, *Pseudococcus adonidum* L. in South Australia', *Australian J. Agric. Res.* **10,** 322–39. *165*

(1962) 'The environments of animals and plants', *J. theoret. Biol.* **2,** 63–8. *20, 129*

(1963) *Animal Populations,* Hutchinson: London. *129*

BRUES, C. T. (1939) 'Studies on the fauna of some thermal springs in Dutch East Indies', *Proc. Amer. Acad. Arts & Sc.,* **73,** 71–95. *100*

BULL, L. B. and MULES, M. W. (1944) 'An investigation of *Myxomatosis cuniculi* with special reference to the possible use of the disease to control rabbit populations in Australia', *J. Counc. Sci. Ind. Res.,* **17,** 79–93. *69*

BULLOUGH, W. S. (1951) *Vertebrate Sexual Cycles,* London: Methuen & Co. *128*

BURNETT, T. (1949) 'The effect of temperature on an insect host-parasite population', *Ecology* **30,** 113–34. *64*

CALABY, J. H. (1963) 'Australia's threatened mammals', *Wildlife* **1,** 15–18. *48*

CARRICK, R. (personal communication), *40*

CARRICKER, M.R. (1951) 'Ecological observations on the distribution of oyster larvae in New Jersey estuaries', *Ecol. Monographs* **21,** 19–38. *184*

CHITTY, D. (1952) 'Mortality among voles (*Microtus agrestis*) at Lake Vyrnwy, Montgomeryshire in 1936–9', *Phil. Trans. R. Soc. Lond. B* **236,** 505–52. *77*

(1957) 'Self-regulation in numbers through changes in viability', *Symp. quant. Biol.* **22,** 277–80.

(1959) 'A note on shock disease', *Ecology* **40,** 728–31. *77*

(1960) 'Population processes in the vole and their relevance to general theory', *Can. J. Zool.* **38,** 99–113 *80.*

(1964) 'Animal numbers and behaviour', in *Fish and Wildlife; a memorial to W. J. K. Harkness* (Ed. J. R. DYMOND, Longmans: Canada), 41–53. *77*

(1965) 'Predicting qualitative changes in insect populations', *Proc. 12th Internat. Congr. Entom. Lond.* 384–6. *82*

(1967) 'The natural selection of self-regulatory behaviour in animal populations', *Proc. Ecol. Soc. Australia* **2,** 51–78. *76, 156*

CHITTY, D., PIMENTEL, D. and KREBS, C. J. (1968) 'Food supply of overwintered voles', *J. Anim. Ecol.* **37,** 113–20. *82*

CHITTY, H. (1961) 'Variations in the weight of the adrenals of the field vole, *Microtus agrestis*', *J. Endocrin.* **22,** 387–93. *80*

CHRISTIAN, J. J. (1950) 'The adreno-pituitary system and population cycles in mammals', *J. Mammal.* **31,** 247–59. *77*

(1955a) 'Effect of population size on the adrenal glands and reproductive organs of male mice in populations of fixed size', *Am. J. Physiol.* **182,** 292–300. *78*

(1955b) 'Effect of population size on the weights of the reproductive organs of white mice', *Am. J. Physiol.* **181,** 477–80. *78*

(1956) 'Adrenal and reproductive responses to population size in mice from freely growing populations', *Ecology* **37,** 258–77. *78*

(1957) 'A review of the endocrine responses in rats and mice to increasing population size including delayed effects on offspring', *Naval Med. Res. Inst. Lect. Rev.* Ser. No. 57–2: 443–62. *78*

(1959) 'The roles of endocrine and behavioural factors in the growth of mammalian populations', *Comparative Endocrinology*, ed. A. GORBMAN, 71–97. *78*

(1961) 'Phenomena associated with population density', *Proc. Natl. Acad. Sci. U.S.* **47,** 428–9. *78*

CHRISTIAN, J. J. and DAVIS, D. E., (1964) 'Endocrines, behaviour and population', *Science, N.Y.* **146,** 1550–60. *78*

CHRISTIAN, J. J. and LEMUNYAN, C. D. (1958) 'Adverse effects of crowding on lactation and reproduction of mice and two generations of their progeny', *Endocrinology* **63,** 517–29. *78*

CLARK, L. R. (1947) 'An ecological study of the Australian plague locust *Chortoicetes terminifera* Walk. in the Bogan-Macquarie outbreak area in N.S.W.', *Bull. Counc. Sci. Indust. Res. Australia,* No. 226. *98*

(1949) 'Behaviour of swarms of hoppers of the Australian plague locust *Chortoicetes terminifera* Walk.', *Bull. Counc. Sci. Indust. Res. Australia,* No. 245. *98*

CLARK, L. R., GEIER, P. W., HUGHES, R. D., and MORRIS, R. F. (1967) *The Ecology of Insect Populations in Theory and Practice,* London: Methuen. *21, 162*

CLARKE G. L. (1954) *Elements of Ecology*, New York: Wiley. *16, 184*

CLARKE, G. M. (1969) *Statistics and Experimental Design*, London: Edward Arnold. *16*

CLARKE, J. R. (1953) 'The effect of fighting on the adrenals, thymus and spleen of the vole *Microtus agrestis*', *J. Endocrin.* **9,** 114–26. *78*

CLAUSEN, C. P. (1951) 'The time factor in biological control', *J. Ecol. Entomol.* **44,** 1–9. *64*

COAD, B. R. (1931) 'Insects captured by airplane are found at surprising heights', *Year book U.S. Dept. Agr.* 1931, 320–3. *181*

COCKERELL, T. D. A. (1934) '"Mimicry" among insects', *Nature, Lond.* **133,** 329–30. *31, 57*

COLE, L. C. (1960a) 'Competitive exclusion', *Science, N.Y.* **132,** 348–9. *50*

(1960b) 'Further competitive exclusion', *Science, N.Y.* **132,** 1675–6. *50*

CRAGG, J. B. (1955) 'The natural history of sheep blowflies in Britain', *Ann. appl. Biol.* **42,** 197–207. *135*

CRIDDLE, N. (1930) 'Some natural factors governing the fluctuations of grouse in Manitoba', *Canadian Field Nat.* **44,** 77–80. *28*

CROGHAN, P. C. (1958a) 'The survival of *Artemia salina* (L.) in various media', *J. exp. Biol.* **35,** 213–18. *102, 123*

(1958b) 'The osmotic and ionic regulation in *Artemia salina* (L.)', *J. exp. Biol.* **35,** 219–33. *123*

(1958c) 'The mechanism of osmotic regulation in *Artemia salina* (L.); the physiology of the branchiae', *J. exp. Biol.* **35,** 234–42. *123*

(1958d) 'The mechanism of osmotic regulation in *Artemia salina* (L.); the physiology of the gut', *J. exp. Biol.* **35,** 243–9. *123*

DAVIDSON, J. (1941) 'Wheat storage problems in South Australia', *J. Agric. S.A.* **44,** 123–36, 243–7, 346, 352, 391–5. *134*

(1944) 'On the relationship between temperature and the rate of development of insects at constant temperatures', *J. Anim. Ecol.*, **13,** 26–38. *87, 229*

DAVIDSON, J. and ANDREWARTHA, H. G. (1948a) 'Annual trends in a natural population of *Thrips imaginis* (Thysanoptera)', *J. Anim. Ecol.* **17,** 193–9. *28, 82, 136, 167*

(1948b) 'The influence of rainfall, evaporation and atmospheric temperature on fluctuations in size of a natural population of *Thrips imaginis* (Thysanoptera)', *J. Anim. Ecol.* **17,** 200–22. *28, 82, 136, 167*

DEEVEY, E. S. JR. (1956) 'The human crop', *Scient. Am.* **194,** 105–12. *29*

DICE, L. R. (1952) *Natural Communities*, Ann Arbor: University of Michigan Press. *16*

DICKSON, R. C. (1949) 'Factors governing the induction of diapause in the oriental fruit moth', *Ann. Entom. Soc. Am.* **42,** 511–37. *126*

DILL, D. B. (1938) *Life, Heat and Altitude*, Cambridge, Mass.: Harvard Univ. Press. *110*

DOBZHANSKY, TH., and WRIGHT, S. (1943) 'Genetics of natural populations. X. Dispersion rates in *Drosophila pseudoobscura*', *Genetics*, **28,** 304–40. *185*

(1947) 'Genetics of natural populations. XV. Rate of diffusion of a mutant gene through a population of *Drosophila pseudoobscura*', *Genetics* **32,** 303–24. *185, 211, 219*

DODD, A. P. (1936) 'The control and eradication of prickly pear in Australia', *Bull. Entomol. Res.* **27,** 503–17. *30*

DORST, H. E. and DAVIS, E. W. (1937) 'Tracing long distance movements of the beet leafhopper in the desert', *J. Econ. Entomol.* **30,** 948–54. *183*

DOUDOROFF, P. (1938) 'Reactions of marine fishes to temperature gradients', *Biol. Bull.* **75,** 494–509. *98*

DOWDESWELL, W. H. (1952) *Animal Ecology*, London: Methuen. *16*

DOWDESWELL, W. H., FISHER, R. A. and FORD, E. B. (1940) 'The quantitative study of populations in the Lepidoptera. I. *Polyommatus icarus*', *Ann. Eugenics* **10,** 123–36. *173*

DUFFY, E. (1956) 'Aerial dispersal in a known spider population', *J. Anim. Ecol.* **5,** 285–111. *182*

DYMOND, J. R. (1947) 'Fluctuations in animal populations with special reference to those of Canada', *Trans. R. Soc. Canada* **41,** 1–34. *49*

EDNEY, E. B. (1947) 'Laboratory studies on the bionomics of the rat fleas *Xenopsylla brasiliensis* Baker and *X. cheopis* Roths. II. Water relations during the cocoon period', *Bull. Entomol. Res.* **38,** 263–80. *123*

ELTON, C. (1927) *Animal Ecology*, London: Sidgwick and Jackson. *3, 16, 48, 190*

(1938) 'Animal numbers and adaptation'. *Evolution: Essays on Aspects of Evolutionary Biology*, Beer, G. R. de ed., Oxford: University Press, 127–37. *57*

(1949) 'Population interspersion: an essay on animal community patterns', *J. Ecol.* **37,** 1–23. *4, 16, 144, 157, 178*

(1958) *The Ecology of Invasions*, London: Methuen. *56, 189*

ELTON, C., and MILLER, R. S. (1954) 'An ecological survey of animal communities: with a practical system of classifying habitats by structural characters', *J. Ecol.* **42,** 460–96. *16*

ERRINGTON, P. L. (1943) 'An analysis of mink predation upon muskrats in north-central United States', *Research Bull. Iowa Agr. Exp. Sta.* **320,** 797–924. *68, 170*

(1944) 'Ecology of the muskrat', *Rep. Iowa. Agr. Exp. Sta.*, 1944, 187–9. *49*

(1945) 'Some contributions of a fifteen-year local study of the northern bob white to a knowledge of population phenomena', *Ecol. Monogr.* **15,** 1–34. *170*

(1946) 'Predation and vertebrate populations', *Quart. Rev. Biol.* **21,** 145–77 and 221–45. *68, 170*

(1948) 'Environmental control for increasing muskrat production', *Trans. 13th North Am. Wildlife Conference*, 596–609. *49*

EVANS, F. C. (1949) 'A population study of the house mouse (*Mus musculus*) following a period of local abundance', *J. Mammal.* **30,** 351–63. *173*

EVANS, J. W. (1933) 'Thrips investigation I: The seasonal fluctuations in numbers of *Thrips imaginis* Bagnall and associated blossom thrips', *J. Coun. Sci. Ind. Res.* (Aust.) **6,** 145–59. *167*

EWER, R. F. (1968) *The Ethology of Mammals*, London: Logos. *46, 59, 103, 120*

FELT, E. P. (1925) 'The dissemination of insects by air currents', *J. Econ. Entomol.* **18,** 152–6. *183*

FENNER F. and RATCLIFFE, F. N. (1965) *Myxomatosis*, Cambridge: University Press. *69*

FINNEY, D. J. (1947) *Probit analysis; a Statistical Treatment of the Sigmoid Response Curve*, Cambridge: University Press. *91, 229, 234*

FISHER, R. A. and FORD, E. B. (1947) 'The spread of a gene in natural conditions in a colony of the moth *Panaxia dominula* (L.)', *Heredity*, **1,** 143–74. *143*

FISHER, R. A. and YATES, F. (1948) *Statistical Tables for Biological, Agricultural, and Medical Research Workers*, London: Oliver and Boyd. *91, 253*

FLANDERS, S. E. (1947) 'Elements of host discovery exemplified by parasitic Hymenoptera', *Ecology* **28,** 299–309. *30, 189*

FORD, E. B. (1945) *Butterflies*, London; Collins. *11, 22*

FRAENKEL, G. and BLEWETT, M. (1944) 'The utilization of metabolic water in insects', *Bull. Entomol. Res.* **35,** 127–37. *103, 115*

FRAENKEL, G. and GUNN, D. L. (1940) *The Orientation of Animals; Kineses, Taxes and Compass Reactions,* Oxford: Clarendon Press. *108*

FRITH, H. J. (1959a) 'The ecology of wild ducks in inland New South Wales I. Waterfowl habitats', *C.S.I.R.O. Wildlife Res.* **4,** 97–107. *158*

(1959b) 'The ecology of wild ducks in inland New South Wales II. Movements', *C.S.I.R.O. Wildlife Res.* **4,** 108–30. *158*

(1959c) 'The ecology of wild ducks in inland New South Wales III. Food habits', *C.S.I.R.O. Wildlife Res.* **4,** 131–55. *158*

(1959d) 'Ecology of wild ducks in inland New South Wales IV. Breeding', *C.S.I.R.O. Wildlife Res.* **4,** 156–81. *158*

(1962) 'Movements of the grey teal *Anas gibberifrons* Muller (Anatidae)', *C.S.I.R.O. Wildlife Res.* **7,** 50–70. *158*

(1963) 'Movements and mortality rates of the black duck and grey teal in south-eastern Australia', *C.S.I.R.O. Wildlife Res.* **8,** 119–31. *158*

(1967) *Waterfowl in Australia,* Sydney: Angus and Robertson. *82*

FRY, F. E. J. (1947) 'Effects of the environment on animal activity', *Univ. Tor. Stud. Biol.* **55,** 1–62. *96*

FRY, F. E. J., BRETT, J. R. and CLAUSEN, G. H. (1942) 'Lethal limits of temperature for young goldfish', *Rev. Canad. Biol.* **1,** 50–6. *96*

FRY, F. E. J., HART, J. S. and WALKER, K. F. (1946) 'Lethal temperature relations for a sample of young speckled trout *Salvelinus fontinalis*', *Publ. Ontario Fish Res. Lab.* **66,** 5–35. *96*

GEIER, P. W. (1963) 'The life history of the codlin moth *Cydia pomonella* (L.) Lepidoptera: Tortricidae, in the Australian Capital Territory', *Australian J. Zool.* **11,** 323–67. *39*

GERKING, S. D. (1952) 'Vital statistics of the fish population of Gordy Lake, Indiana', *Trans. Am. Fish. Soc.* **82,** 48–67. *178*

GLOVER, P. E., JACKSON, C. H. N., ROBERTSON, A. G. and THOMSON, W. E. F. (1955) 'The extermination of the tse-tse fly, *Glossina morsitans* West., at Abercorn, Northern Rhodesia', *Bull. Entomol. Res.* **46,** 57–67. *11, 56*

GOWERS, E. (1948) *Plain Words,* London: H.M. Stat. Office. *15*

GREEN, R. G. and EVANS, C. A. (1940) 'Studies on a population cycle of snowshoe hares on the Lake Alexander area', *J. Wildlife Management* **4,** 220–38. *77*

GREEN, R. G. and LARSEN, C. L. (1938) 'A description of shock disease in the snowshoe hare', *Am. J. Hyg.* **28,** 190–212. *77*

HALL, F. G. (1922) 'Vital limits of exsiccation of certain animals', *Biol. Bull.* **42,** 31–51. *102*

HARDIN, G. (1956) 'Meaninglessness of the word protoplasm', *Sci. Monthly* **82,** 112–20. *46*

(1960) 'The competitive exclusion principle', *Science, N.Y.* **131,** 1292–8. *50*

HARKER, J. E. (1956) 'Factors controlling the diurnal rhythm of activity of *Periplaneta americana* L.', *J. exp. Biol.* **33,** 224–34. *124*

HEWITT, C. G. (1921) *The Conservation of the Wildlife of Canada,* New York: Scribner. *48*

HINTON, H. E. (1960) 'Cryptobiosis in the larva of *Polypedilum vanderplanki* Hint. (Chironomidae)', *J. Insect Physiol.* **5,** 286–300. *101*

HOCHBAUM, H. A. (1955) *Travels and Traditions of Waterfowl,* Minneapolis: Univ. Minnesota Press. *184*

HOWARD, L. O. (1931) *The Insect Menace,* New York: Century Press. *143*

HOWARD, L. O. and FISKE, W. F. (1911) 'The importation into the United States of the parasites of the gipsy moth and the brown-tail moth', *Bull. U.S. Bur. Entom.* No. 91. *156, 160*

HOY, J. M. (1954) 'A new species of *Eriococcus* Targ. (Hemiptera, Coccidae) attacking *Leptospermum* in New Zealand', *Trans. R. Soc. N.Z.* **82,** 465–74. *182*

HUNTSMAN, A. G. (1946) 'Heat stroke in Canadian maritime stream fishes', *J. Fish Res. Bd Canada* **6,** 476–82. *96*

HUTCHINSON, G. E. and DEEVEY, E. S. JR. (1949) 'Ecological studies on populations', *Survey Biol. Prog.* **1,** 325–59. *135*

JACKSON, C. H. N. (1937) 'Some new methods in the study of *Glossina morsitans*', *Proc. Zool. Soc. Lond.* 1936, 811–96. *25*

(1939) 'The analysis of an animal population', *J. Anim. Ecol.* **8,** 238–46. *173.*

(1940) 'The analysis of a tse-tse fly population', *Ann. Eugenics* **10,** 322-69. *178*

(1944) 'The analysis of a tse-tse fly population II', *Ann. Eugenics* **12,** 176–205. *178*

(1948) 'The analysis of a tse-tse fly population III', *Ann. Eugenics* **14,** 91–108. *178, 202*

JOHNSON, C. G. (1940) 'The longevity of the fasting bed-bug (*Cimex lectularius*) under experimental conditions and particularly in relation to the saturation deficiency law of water loss', *Parasitology* **32,** 329–70. *91*

(1950a) 'The comparison of suction trap, sticky trap and tow net for the quantitative sampling of small airborne insects', *Ann. appl. Biol.* **37,** 268–85. *166*

(1950b) 'Infestation of a bean field by *Aphis fabae* Scop. in relation to wind direction', *Ann. appl. Biol.* **37,** 441–50. *181*

(1952a) 'The changing numbers of *Aphis fabae* Scop. flying at crop level, in relation to current weather and to the population on the crop' *Ann. appl. Biol.* **39,** 525–47. *166, 181*

(1952b) 'A new approach to the problem of the spread of aphids and to insect trapping', *Nature Lond.* **170,** 147–8. *166, 181*

(1954) 'Aphid migration in relation to weather', *Biol. Rev.* **29,** 87–118. *118*

(1955) 'Ecological aspects of aphid flight and dispersal', *Rep. Rothamsted Exp. Sta.*, 1955, 191–201. *166, 181*

(1956) 'Changing views on aphid dispersal', *N.A.A.S. Quart. Rev.*, No. 32. *181*

(1957) 'The vertical distribution of aphids in the air and the temperature lapse rate', *Qt. Jl. R. Met. Soc.* **83,** 194–201. *181*

JOHNSON, C. G., and TAYLOR, L. R. (1957) 'Periodism and energy summation with special reference to flight rhythms in aphids', *J. exp. Biol.* **34,** 209–21. *181*

KLOMP, H. (1966a) 'The dynamics of a field population of the pine looper *Bupalis piniarus*', *Advan. Ecol. Res.* **3,** 207–305.

(1966b) 'The interrelations of some approaches to the concept of density dependence in animal populations', *Med. Landbow. Wageninen*, Nederland **66,** 1–10.

KLUIJVER, H. N. (1951) 'The population ecology of the great tit *Parus m. major*', *Ardea* **39,** 1–135. *35, 41, 145, 170*

KNIPLING, E. F. (1955) 'Possibilities of insect control or eradication through the use of sexually sterile males', *J. econ. Entom.* **48,** 459–62. *56*

(1960) 'The eradication of the screwworm fly *Caliroa hominovorax*', *Scient. Am.* **203,** 54–61. *57*

KOCH, H. (1938) 'The absorption of chloride ions by the anal papillae of diptera larvae', *J. exp. Biol.* **15,** 152–60. *111*

KOZHANCHIKOV, I. V. (1938) 'Physiological conditions of cold-hardiness in insects', *Bull. Entom. Res.* **29,** 253–62. *92*

KREBS, C. J. (1964) 'The lemming cycle at Baker Lake, Northwest Territories, during 1959–62', *Arctic Inst. N. Am. Tech. Paper* No. 15. *79*

(1966) 'Demographic changes in fluctuating populations of *Microtus californicus*', *Ecol. Monogr.* **36,** 239–73. *79*

KROGH, A. (1937) 'Osmotic regulation in fresh water fishes by active absorption of chloride ions', *Z. Vergl. Physiol.* **24,** 656–66. *111*

(1939) *Osmotic regulation in aquatic animals*, Cambridge: University Press. *110, 123*

LEEPER, G. W. (1941) 'The scientist's English', *Australian J. Sci.* **3,** 121–3. *15*

(1952) 'An index of ease of reading', *Australian J. Sci.* **15,** 31–2. *15*

LEES, A. D. (1946) 'The water balance in *Ixodes ricinus* L. and certain other species of ticks', *Parasitology* **37,** 1–20. *113*

(1947) 'Transpiration and the structure of the epicuticle in ticks', *J. exp. Biol.* **23,** 379–410. *113, 123*

LEOPOLD, A. (1933) *Game Management*, New York: Charles Scribners Sons. *135*

(1943) 'Deer irruptions', *Wisconsin Conserv. Bull.*, August, 1943. *34*

LESLIE, P. H. (1952) 'The estimation of population parameters from data obtained by means of the capture-recapture-recapture method', *Biometrika* **39,** 363–8. *178*

LESLIE, P. H. and CHITTY. D. (1951) 'The estimation of population parameters from data obtained from the capture-recapture method. I. The maximum likelihood equations for estimating death-rate', *Biometrika* **38,** 269–92. *178*

LINDAUER, M. (1961) *Communication among Social Bees*, Cambridge, Mass.: Harvard University Press. *136*

LLOYD, L. (1937) 'Observations on sewage flies; their seasonal incidence and abundance', *J. Inst. Sewage Purification* (1937), 1–16. *140*

(1941) 'The seasonal rhythm of a fly (*Spaniotoma minima*) and some theoretical considerations', *Trans. R. Soc. Trop. Med. Hyg.* **35,** 93–104. *140, 171*

(1943) 'Materials for a study in animal competition. II. The fauna of the sewage beds. III. The seasonal rhythm of *Psychoda alternata* Say, and an effect of intraspecific competition', *Ann. appl. Biol.* **30,** 47–60. *140*

LLOYD, L., GRAHAM, J. F. and REYNOLDSON, T. B. (1940) 'Materials for study, in animal competition. The fauna of the sewage bacteria beds', *Ann. appl. Biol.* **27,** 122–50. *140*

MCCABE, T. T. and BLANCHARD, B. D. (1950) *Three Species of Peromyscus*, Santa Barbara, California: Rood. *13*

MCDONALD, G. (1957) *Epidemiology and Control of Malaria*, London: Oxford University Press. *72*

MACAN, T. T. and WORTHINGTON, E. B. (1951) *Life in Lakes and Rivers*, London: Collins. *22*

MACFADYEN, A. (1957) *Animal Ecology; Aims and Methods*, London: Pitman. *16*

MACKENZIE, J. M. D. (1952) 'Fluctuations in the numbers of British tetraonids', *J. Anim. Ecol.* **21,** 128–53. *17*

MACKERRAS, I. M. and MACKERRAS, M. J. (1944) 'Sheep blowfly investigations. The attractiveness of sheep for *Lucilia sericata*', *Bull. Counc. Sci. Indust. Research Australia*, No. 181.

MACMILLEN, R. E. and LEE, A. K. (1969) 'Water metabolism of Australian hopping mice', *Comp. Biochem. Physiol.* **28,** 493–514. *118, 123*

MADGE, P. E. (1954) 'Control of the underground grass caterpillar', *J. Dept. Agric. S. Australia*, March 1954. *171*

MAELZER, D. A. (1958) 'The ecology of *Aphodius howitti*', Thesis (unpublished) in library of University of Adelaide. *107*

(1965) 'A discussion of components of environment in ecology', *J. theoret. Biol.* **8,** 141–62. *19*

MAIL, G. A. (1930) 'Winter soil temperatures and their relation to subterranean insect survival', *J. agric. Res.* **41,** 571–92. *52*

MAIL, G. A. and SALT, R. W. (1933) 'Temperature as a possible limiting factor in the northern spread of the Colorado potato beetle', *J. econ. Entomol.* **26,** 1068–75. *52*

MARSDEN, H. M. and HOLLER, N. R. (1964) 'Social behaviour in confined populations of cottontail and the swamp rabbit', *Wildlife Monogr.* Chestertown No. 13.

MARSHALL, F. H. A. (1942) 'Exteroceptive factors in sexual periodicity', *Biol. Rev.* **17,** 68–89. *127*

MATHER, K. (1943) *Statistical Analysisin Biology*, Londcn: Methuen.

MATHEWS, G. V. T. (1955) *Bird Navigation*, Cambridge: University Press. *184*

MATTHEE, J. J. (1951) 'The structure and physiology of the egg of *Locustana pardalina* (Walk)', *Scient. Bull. Dept. Agr. Sth. Africa*, No. 316. *112*

MAUGHAM, W. S. (1948) *The Summing Up*, London: Heinemann. *14*

MEAD-BRIGGS, A. R. and RUDGE, A. J. B. (1960) 'Breeding of the rabbit flea *Spilopsyllus cuniculi* (Dale): requirement of a "factor" from a pregnant rabbit for ovarian maturation', *Nature, Lond.* **187,** 1136. *23*

MICHELBACHER, A. E. and SMITH, R. L. (1943) 'Some natural factors limiting the abundance of the alfalfa butterfly', *Hilgardia* **12;** 366–97. *74*

MILNE, A. (1943) 'The comparison of sheep tick populations (*Ixodes ricinus* L.)', *Ann. appl. Biol.* **30,** 240–50. *172*

(1949) 'The ecology of the sheep tick *Ixodes ricinus* L. Host relationships of the tick. 2. Observations on hill and moorland grazings in northern England', *Parasitology* **39,** 173–97. *26*

(1950) 'The ecology of the sheep tick *Ixodes ricinus* L., Spatial distribution' *Parasitology* **40,** 35–45. *55*

(1951) 'The seasonal and diurnal activities of individual sheep ticks (*Ixodes ricinus* L.)', *Parasitology* **41,** 189–208. *26, 55*

(1952) 'Features of the ecology and control of the sheep tick, *Ixodes ricinus* L. in Britain', *Ann. appl. Biol.* **39,** 144–6. *55*

MOORE, J. A. (1939) 'Temperature tolerance and rates of development in the eggs of Amphibia', *Ecology* **20,** 459–78. *99*

(1942) 'The role of temperature in the speciation of frogs', *Biol. Symp.* **6,** 189–213. *99*

MORONEY, M. J. (1953) *Facts from Figures*, London: Penguin Books. *16, 223*

MORRIS, R. F. (1959) 'Single-factor analysis in population dynamics,' *Ecology* **40,** 580–8.

(1963) 'Predictive population equations', *Mem. Entom. Soc. Canada* **32,** 16–21. *21*

MYERS, K. and PARKER, B. S. (1965) 'A study of the wild rabbit in climatically different regions in eastern Australia: I. Patterns of distribution', *C.S.I.R.O. Wildlife Res.* **10,** 33–72. *53*

MYERS, K. and POOLE, W. E. (1959) 'A study of the biology of the wild rabbit in confined populations: I. The effect of density on home range and the formation of breeding groups', *C.S.I.R.O. Wildlife Res.* **4,** 14–26. *42*

(1961) 'A study of the biology of the wild rabbit in confined populations: II. The effects of season and population growth on behaviour', *C.S.I.R.O. Wildlife Res.* **6,** 1–41. *42*

(1962) 'A study of the biology of the wild rabbit in confined populations: III. Reproduction', *Australian J. Zool.* **10,** 225–67. *42*

(1963a) 'A study of the biology of the wild rabbit in confined populations: IV. The effect of rabbit grazing on sown pastures', *J. Ecol.* **51,** 435–51. *42*

(1963b) 'A study of the biology of the wild rabbit in confined populations: V. Population dynamics', *C.S.I.R.O. Wildlife Res.* **8,** 166–203. *42*

MYKYTOWYCZ, R. (1958) 'Social behaviour of an experimental colony of wild rabbits, *Oryctolagus cuniculus* (L): I. Establishment of the colony', *C.S.I.R.O. Wildlife Res.* **3,** 7–25. *42*

(1959) 'Social behaviour of an experimental colony of wild rabbits, *Oryctolagus cuniculus* (L): II. First breeding season', *C.S.I.R.O. Wildlife Res.* **4,** 1–13. *42*

(1960) 'Social behaviour in an experimental colony of wild rabbits *Oryctolagus cuniculus* (L): III. Second breeding season', *C.S.I.R.O. Wildlife Res.* **5,** 1–20. *42*

(1961) 'Social behaviour in an experimental colony of wild rabbits *Oryctolagus cuniculus* (L): IV. Conclusions, outbreak of myxomatosis, third breeding season and starvation', *C.S.I.R.O. Wildlife Res.* **6,** 142–55. *42*

(1968) 'Territorial marking by wild rabbits', *Scient. Am.* **218,** 116–19. *42*

NEWSOME, A. E. (1967) 'A simple biological method of measuring the food supply of house mice', *J. Anim. Ecol.* **36,** 645–50. *82*

NEWSON, R. (1963) 'Differences in numbers, reproduction and survival between two neighbouring populations of bank voles (*Clethrionomys glareolus*)', *Ecology* **43,** 733–8. *79*

NICHOLSON, A. J. (1947) 'Fluctuation of animal populations', *Rep. 26th Meeting A.N.Z.A.A.S.*, Perth, 1947. *190*

(1954) 'An outline of the dynamics of animal populations', *Australian J. Zool.* **2,** 9–65. *21, 157, 162*

(1958) 'Dynamics of insect populations', *Ann. Rev. Entom.* **3,** 107–36. *157*

NIVEN, S. (1967) 'The stochastic simulation of *Tribolium* populations', *Physiol. Zool.* **40,** 67–82. *83*

ODUM, E. P. (1953) *Fundamentals of Ecology*, Philadelphia: Saunders. *59*

ORWELL, G. (1946) 'The English language', *J. and Proc. Aust. Chem. Inst.*, Sept. 1946. *16*

PARK, T. (1948) 'Experimental studies of interspecies competition. I. Competition between populations of flour beetles *Tribolium confusum* Duval and *Tribolium castaneum* Herbst', *Ecol. Monogr.* **18,** 265–308. *83*

(1954) 'Experimental studies of interspecies competition. II. Temperature, humidity and competition in two species of *Tribolium*', *Physiol. Zool.* **27,** 177–238. *83*

PAYNE, N. M. (1927) 'Measures of insect cold-hardiness', *Biol. Bull.* **52,** 449–57. *102*

PEARSON, O. P. (1947) 'The rate of metabolism of some small mammals', *Ecology* **28,** 127–45. *135*

PETRUSEWICZ, K. (1957) 'Investigation of experimentally induced population growth', *Ekol. Polska A.* **5,** 281–309. *78*

PHILLIPSON, J. (1966) *Ecological Energetics*, London: Edward Arnold. *16*

PITELKA, F. A. (1958) 'Some aspects of population structure in the short-term cycle of the brown lemming in northern Alaska', *Cold Spring. Harb. Symp. quant. Biol.* **22,** 237–51. *79*

POLLISTER, A. W. and MOORE, J. A. (1937) 'Tables for the normal development of *Rana sylvatica*', *Anat. Rec.* **68,** 489–96. *99*

POMEROY, D. E. (1966) The ecology of *Helicella virgata* and related species in South Australia, Thesis, University of Adelaide Library. *116*

POPPER, K. (1959) *Logic of Scientific Discovery*, London: Hutchinson. *158*

POTTS, W. H. and JACKSON, C. H. N. (1953) 'The Shinyanga game-destruction experiment', *Bull. entomol. Res.* **43,** 365–74. *26*

POWSNER, L. (1935) 'On the effects of temperature on the duration of the development stages of *Drosophila melanogaster*', *Physiol. Zool.* **8,** 474–520. *87*

RAINEY, R. C. and WALOFF, Z. (1948) 'Desert locust migrations and synoptic meteorology in the Gulf of Aden area', *J. Anim. Ecol.* **17,** 110–12. *183*

(1951) 'Flying locusts and convection currents', *Bull. Anti-loc. Res. Centre,* London **9,** 51–70. *183*

RAMSAY, J. A. (1952) *Physiological Approach to the Lower Animals,* Cambridge: University Press. *110*

REYNOLDSON, T. B. (1966) 'The distribution and abundance of lake-dwelling triclads – towards a hypothesis', *Advan. Ecol. Res.* **3,** 1–71. *47*

RICHARDS, L. A. (1949) 'Methods of measuring soil moisture tension', *Soil Sci.* **68,** 95–112. *107*

RITCHIE, J. (1920) *The Influence of Man on Life in Scotland,* Cambridge: University Press. *48*

ROTH, L. M. and WILLIS, E. R. (1951) 'The effects of desiccation and starvation on the humidity behaviour and water balance of *Tribolium confusum* and *Tribolium castaneum*', *J. exp. Zool.* **118,** 337–61. *103*

ROWAN, W. (1938) 'Light and seasonal reproduction in animals', *Biol. Rev.* **13,** 374–402. *127*

RUSSELL, B. (1956) *Portraits from Memory,* London: George Allen and Unwin.

RUSSELL, F. S. (1939) 'Hydrographical and biological conditions in the North Sea as indicated by Plankton organisms', *J. Cons. Int. Explor. Mer.* **14,** 171–92. *184*

SALT, G., HOLLICK, F. S. J., RAW, F. and BRIAN, M. V. (1948) 'The arthropod population of pasture soil', *J. Anim. Ecol.* **17,** 139–50. *171*

SALT, R. W. (1950) 'The time as a factor in the freezing of undercooled insects', *Can. J. Res.* D. **25,** 66–86. *94*

SANARELLI, G. (1898) 'Das myxomatogene Virus. Beitrag zur Studium der Krankheitserreger ausserhalb des Sichtbaren', *Zbl. Bakt.* (*Abt.* 1) **23,** 865. *69*

SCHMIDT, J. (1925) 'The breeding places of the eel', *Smithsonian Inst. Ann. Rep.* 1924, 279–316. *184*

SCHMIDT, P. (1918) 'Anabiosis of the earthworm', *J. exp. Zool.* **27,** 57–72. *102*

SCHMIDT-NIELSEN, B. and SCHMIDT-NIELSEN, K. (1949) 'The water economy of desert mammals', *Sci. Monthly* **69,** 180–5. *110*

(1950) 'Evaporative water loss in desert rodents in their natural habitat', *Ecology* **31,** 75–85. *110*

(1951) 'A complete account of the water metabolism in the kangaroo rat and an experimental verification', *J. Cell. Comp. Physiol.* **38,** 165–81. *109, 120*

SCHMIDT-NIELSEN, B., SCHMIDT-NIELSEN, K. and JARNUM, S. A. (1956) 'Water balance of the camel', *Amer. J. Physiol.* **185,** 185–94. *103, 120*

SCHMIDT-NIELSEN, K. (1964) *Desert Animals*, London: Oxford University Press. *102, 121, 123*

SCHMIDT-NIELSEN, K. and SCHMIDT-NIELSEN, B. (1952) 'Water metabolism of desert mammals', *Physiol. Rev.* **32,** 135–66. *103, 118*

SCHMIDT-NIELSEN, K. and SLADEN, W. J. L. (1958) 'Nasal salt secretion in the Humboldt penguin', *Nature, Lond.* **181,** 1217–18. *123*

SCHMIEDER, R. G. (1933) 'The polymorphic forms of *Melittobia chalybii* Ashm. and the determining factors involved in their production', *Biol. Bull.* Woods Hole **65,** 338–54. *23*

(1939) 'The significance of two types of larvae in *Sphecophaga burra* (Cresson) and the factors conditioning them', *Entom. News* **50,** 125–31. *23*

SHELFORD, V. E. (1927) 'An experimental study of the relations of the codling moth to weather and climate', *Bull. Illinois Nat. Hist. Surv.* **16,** 311–440. *92*

SMITH, F. E. (1961) 'Density dependence in the Australian thrips', *Ecology* **42,** 403–7.

SMITH, H. S. (1935) 'The role of biotic factors in determining population densities', *J. Econ. Entomol.* **28,** 873–98. *20, 57, 156, 162*

(1939) 'Insect populations in relation to biological control', *Ecol. Monogr.* **9,** 311–20. *30, 61, 189*

SOLOMON, M. E. (1957) 'Dynamics of insect populations', *Ann. Rev. Entomol.* **2,** 121–42. *157*

(1964) 'Analysis of processes involved in the natural control of insects', *Advan. Ecol. Res.* **2,** 1–58.

SOUTHWICK, C. H. (1955a) 'The population dynamics of confined house mice supplied with unlimited food', *Ecology* **36,** 212–25. *78*

(1955b) 'Regulatory mechanisms of house mouse populations: social behaviour affecting litter survival', *Ecology* **36,** 627–34. *78*

SOUTHWOOD, T. R. E. (1967) 'The interpretation of population change', *J. Anim. Ecol.* **36,** 519–29.

STODDART, E. and MYERS, K. (1966) 'The effects of different foods on confined populations of wild rabbits, *Oryctolagus cuniculus* (L)', *C.S.I.R.O. Wildlife Res.* **11,** 111–24. *23*

TALBOT, L. M. and TALBOT, M. H. (1963) 'The wildebeest in western Masailand', *Wildlife Monogr.* Chestertown No. 12, 1–88. *59*

TANNER, J. T. (1966) 'Effects of population density on growth rates of animal populations', *Ecology* **47,** 733–45.

TAYLOR, L. R. (1957) 'Temperature relations of teneral development and behaviour in *Aphis fabae* Scop.' *J. exp. Biol.* **34,** 189–208. *181*

THOMPSON, C. G. and STEINHAUS, E. A. (1950) 'Further tests using a polyhedrosis virus to control the alfalfa caterpillar', *Hilgardia* **19,** 411–45. *74*

THORSON, T. and SVIHLA, A. (1943) 'Correlation of the habitats of amphibians with their ability to survive the loss of body water', *Ecology* **24,** 374–81. *102*

TOTHILL, J. D., TAYLOR, T. H. C. and PAYNE, R. W. (1930) *The Coconut Moth in Fiji*, London: Imperial Bureau of Entomology. *66*

ULLYETT, G. C. (1950) 'Competition for food and allied phenomena in sheep blowfly populations', *Phil. Trans. R. Soc. Lond. B* **234,** 77–174. *31, 131*

VARLEY, G. C. (1947) 'The natural control of population balance in knapweed gall fly (*Urophora jaceana*)', *J. Anim. Ecol.* **16,** 139–87. *20*

VARLEY, G. C. and GRADWELL, G. R. (1968) 'Population models for the winter moth' *Symp. R. Entomol. Soc. Lond.* **4,** 132–42.

VÔUTE, A. D. (1946) 'Regulation of the density of the insect-populations in virgin forests and cultivated woods', *Arch. Neerl. Zool.* **7,** 435–70. *143*

WALOFF, Z. (1946a) 'A long-range migration of the desert locust from southern Morocco to Portugal, with an analysis of concurrent weather conditions', *Proc. R. Entomol. Soc. Lond., A* **21,** 81–4. *183*

(1946b) 'Seasonal breeding and migrations of the desert locust (*Schistocerca gregaria* Forsk.) in eastern Africa', *Mem. Anti-locust Research Centre London,* No. 1. *58, 183*

WALTERS, S. S. (1966) The conservation of water in starving *Tenebrio molitor* L., Thesis, Univ. of Adelaide Library. *115*

WATERHOUSE, D. F. (1947) 'The relative importance of live sheep and of carrion as breeding grounds for the Australian sheep blowfly *Lucilia cuprina*', *Bull. Counc. Sci. Indust. Res. Australia* No. 217. *33, 131*

WATT, K. E. F. (1964) 'Density dependence in population fluctuations', *Can. Entomol.* **96,** 1147–8.

WEISS, P. (1950) 'Perspectives in the field of morphogenesis', *Quart. Rev. Biol.* **25,** 177–98. *16, 88*

WELLINGTON, W. G. (1949) 'The effects of temperature and moisture upon the behaviour of the spruce budworm, *Choristoneura fumiferana* Clemens (Lepidoptera: Tortricidae) I. The relative importance of graded temperatures and rates of evaporation in producing aggregations of larvae', *Scient. Agric.* **29,** 201–51. *104*

WHITE, T. C. R. (1966) 'Food and outbreaks of phytophagous insects', Thesis, Univ. of Adelaide Library. *23*

(1969) 'An index to measure weather induced stress of trees associated with outbreaks of psyllids in Australia', *Ecology 50,* vol. 5.

WICHTERMAN, R. (1953) *The biology of Paramoecium,* New York: Blakiston. *213*

WIGGLESWORTH, V. B. (1932) 'On the function of the so called rectal glands of insects', *Q. J. microsc. Sci.* **75,** 131–50. *112*

(1933) 'The effect of salt on the anal gills of the mosquito larvae', *J. exp. Biol.* **10,** 1–15. *111, 123*

(1948) 'The insect cuticle', *Biol. Rev.* **23,** 408–51. *112*

WIJNGAARDEN, A. VAN (1960) 'The population dynamics of four confined populations of the continental vole *Microtus arvalis* (Pallas)', *Vers. von landbouk. onderzoek.* **66.22,** 1–28. *79*

WILDERMUTH, L. L. (1911) 'The alfalfa caterpillar (*Eurymus eurytheme*)', *U.S.D.A. Bur. Entomol. Circ.* **133,** 1–4. *74*

WILLIAMS, C. B. (1940) 'An analysis of four years'.captures of insects in a light trap. II. The effect of weather conditions on nsect activity; and the estimation and forecasting of changes in the insect population', *Trans. Ry Entomol. Soc. Lond.* **90,** 227–306. *167*

WILSON, E. O. (1963) 'Chemical communication among animals', *Recent Progress* **19,** 673–716. *58*

WILSON, F. (1961) 'Adult reproductive behaviour in *Asolcus basalis* (Hymenoptera: Scelionidae)', *Australian J. Zool.* **9,** 737–51. *37*

Index